Sendai Virus Vector

Yoshiyuki Nagai
Editor

Sendai Virus Vector

Advantages and Applications

Springer

Editor
Yoshiyuki Nagai
Director
RIKEN Center of Research Network for Infectious Diseases (CRNID)
Jimbocho 101 Bldg. 8th fl.
1-101 Kandajimbo-cho, Chiyoda-ku
Tokyo 101-0051, Japan
yoshi-nagai@riken.jp

ISBN 978-4-431-54555-2 ISBN 978-4-431-54556-9 (eBook)
DOI 10.1007/978-4-431-54556-9
Springer Tokyo Heidelberg New York Dordrecht London

Library of Congress Control Number: 2013956235

Printed on acid-free paper

Springer is part of Springer Science+Business Media (www.springer.com)

Foreword

The present book on Sendai virus vectors describes a fascinating journey from the discovery of the virus in 1953 in Japan to early molecular biological work and its promising use as a weapon against human diseases. The virus was initially called hemagglutinating virus of Japan (HVJ) and much of the work on the virus was subsequently done in Japan. Thus, it is highly appropriate that the majority of the authors for this book hail from Japan, both from academia as well as industry. Yoshiyuki Nagai has been one of the central players in the long journey from the early biological work on Sendai virus to the more recent developments of Sendai virus vectors and he is thus eminently suited to be the editor of this book.

Sendai virus is now recognized to be of murine origin and to have a close cousin in the human parainfluenza virus 1 (HPIV1). The virus belongs to the large family of negative strand RNA viruses of which influenza virus and measles virus are the members causing the most morbidity and mortality in humans. Much of what we know about the function and biology of the genes of negative strand RNA viruses is based on the availability of reverse genetics for this virus family. More than two decades ago my laboratory (Enami et al. 1990) developed a reverse genetics system for influenza viruses allowing the introductions of site-specific mutations into the RNA genome of influenza viruses using cloned DNA (hence the name reverse genetics—one has to go back to DNA to introduce/replace a gene in the RNA genome of the virus). The critical breakthrough of reverse genetics for the non-segmented (negative strand) RNA viruses came from the Conzelmann group in 1994 when they succeeded in obtaining an infectious rabies virus from cloned DNA (Schnell et al. 1994). This was followed by reverse genetics systems for many of the other negative strand RNA viruses including Sendai virus, measles virus, respiratory syncytial virus, and vesicular stomatitis virus.

However, the road to these accomplishments of reverse genetics was not an easy one. For the younger readers of this book, the difficult story of early reports on the rescue of a negative strand RNA virus (measles virus) from a cDNA clone will be recounted. In 1990/1991, two papers appeared that described the extraordinary feat of rescuing infectious measles virus and measles virus mutants after injecting measles virus cDNA into cells. I was actually a reviewer for both papers and I was stunned by the beauty and conciseness of the papers [one published in *The EMBO Journal* (Ballart et al. 1990) and the other one in the *Journal of Virology* (Ballart et al. 1991)]. I could not find any fault in the two manuscripts (and being a competitor in the field it is frequently easy to question data and ask for clarifications, etc.). It turned out, however, that the first author of both papers had manipulated the data. The sequences of the viral mutants shown in the paper were not based on the genomic RNA of rescued virus but just on the plasmid DNA representing the mutant sequences. No infectious RNA virus (measles virus) was ever rescued. The first author was actually the "skilled" one doing the microinjection and the characterization of the measles virus mutants. When other groups, including our own, could not repeat the revolutionary new technology of microinjection, the pressure on the first author must have become stronger and stronger and actually led to his suicide. There are several lessons to be learned from this tragic episode. Yes, competition is important, but it should not drive people to cut corners and make up data. Science is based on the reproducibility and ultimately the truth, and thus tampering with it is simply a no-no. Another lesson was a painful one for me on a personal level. I, as a careful reviewer, could not see any problems with the manuscript; if someone wants to mislead you it can happen easily. Let's be careful and not overinterpret our own data. Checks and balances must be built into our work process. A third lesson was that the results looked very reasonable. I thought that the technology as described could/should have worked, but the data were obviously not there. Moving forward almost two and a half decades later, we now have evidence-based medicine. Just because we think it could work is not enough reason to believe it. We need ample proof in science and medicine.

Progress on Sendai virus vectors described in this book has been phenomenal. Much of the earlier work was just good basic science. For example, the observation by Calain and Roux (1993) that the genome of Sendai viruses has to be a multiple of six was an astounding finding. Without this knowledge, any translational research with Sendai virus vectors would be impossible today. This is a very good example that basic research is absolutely necessary if we want to have something to translate into a useful drug, medication, or treatment modality.

The concepts and technologies described by Akihiro Iida and Makoto Inoue in Chap. 3 of the book are impressive as new generations of Sendai virus vectors are becoming a reality. The safety characteristics of these highly imaginative constructs are convincing, and novel applications of these vectors (not even considered at the present time) should be forthcoming in the future. While the safety concerns with these Sendai virus vectors have been adequately addressed, some of the applications in the vaccine field need to be scrutinized. Why should Sendai virus vectors expressing HIV transgenes tested as HIV vaccines not cause the same adverse reactions as adenovirus-based HIV vaccines? In these HIV vaccine trials, it is thought that people who have an immune recollection against adenoviruses (because they were naturally infected previously) would activate their lymphocytes. These activated lymphocytes, having an enhanced receptor density for HIV, would be more easily infected by HIV. This is thought to result in higher HIV infection rates in vaccinated patients than control non-vaccinated subjects. Since most humans have antibodies against Sendai virus (because of prior infections with cross-reacting parainfluenza virus 1, is it possible that Sendai virus-based HIV vaccines may run into problems similar to those in the earlier adenovirus-based HIV vaccine trials? Sendai virus-based vaccines for other indications, including infections by *Trypanosoma cruzi*, should be free of such possible complications (Chap. 5).

Most exciting are new developments using modified Sendai viruses as unique oncolytic agents. Mutants capable to induce massive cell-to-cell fusion and made more cancer cell-specific by altering the cleavage site of the F protein (to enhance activation of the virus in cancer cells only) should be promising in the future (Chap. 6). It appears that the rules and regulations to do laboratory experiments and clinical trials in Japan and other countries are more rational and user-friendly than in the United States. Thus, it will not be surprising when many of the future breakthroughs in biomedical research will come from ex-USA. The current climate in the United States is clearly anti-progress and hampered by excessive safety requirements imposed by government (FDA, etc.) and local review committees. The result likely is that countries with harmonized and user-friendly review guidelines will be the source of effective new medicines, and other countries which erect one bureaucratic/political hurdle after the other will lag behind in these promising developments.

One of the most inspiring chapters is written by Noemi Fusaki and Hiroshi Ban (Chap. 7). Sendai virus vectors expressing the reprogramming factors OCT3/4, SOX2, KLF4, and c-MYC can now be used to generate induced pluripotent stem cells (iPSCs). Such iPSCs established from patients with intractable genetic diseases will be of extraordinary value to advance research on such diseases and to possibly find drugs alleviating these conditions. Regenerative medicine may enter a new era with these Sendai virus-based vectors. iPSC lines generated by the vectors can then be "repaired" in the affected gene and given back to the patient. The proof of principle for such potential of combining human iPSCs with genetic correction was illustrated by the correction of alpha-1 antitrypsin deficiency (A1ATD) of patient-derived iPSCs and their differentiation into hepatocytes that exhibited normal liver functions in transplanted immunodeficient mice.

A human trial of patients with peripheral arterial disease is described in the last chapter. A non-replicating Sendai virus vector expressing basic fibroblast growth factor (FGF-2) was given to 12 patients with critical limb ischemia with the hope that exogenously expressed FGF-2 would improve vascularization. The results of the first-in-man trial are promising and it is hoped that this vector-based approach will rapidly move into further clinical trials.

The present book on Sendai virus vectors and their applications for human medicine shows us that the future has already arrived. Very soon new and unimagined applications will become available to prevent or treat previously untreatable diseases. The rapid and powerful technique to generate iPSCs using safe and effective Sendai virus vectors is truly a wonderful and promising tool. This book is an inspiring tour de force into novel technologies and possibilities and is highly recommended for students and researchers alike.

New York, USA Peter Palese

References

Ballart I, Eschle D, Cattaneo R, Schmid A, Metzler M, Chan J, Pifko-Hirst S, Udem SA, Billeter MA (1990) Infectious measles virus from cloned cDNA. EMBO J 9:379–384. Retraction in: Eschle D, Cattaneo R, Schmid A, Metzler M, Chan J, Pifko-Hirst S, Udem SA, Billeter MA (1991) EMBO J 10:3558

Ballart I, Huber M, Schmid A, Cattaneo R, Billeter MA (1991) Functional and nonfunctional measles virus matrix genes from lethal human brain infections. J Virol 65:3161–3166. Retraction in: Huber M, Schmid A, Cattaneo R, Billeter MA (1991) J Virol 65:5656

Calain P, Roux L (1993) The rule of six, a basic feature for efficient replication of Sendai virus defective interfering RNA. J Virol 67:4822–4830

Enami M, Luytjes W, Krystal M, Palese P (1990) Introduction of site-specific mutations into the genome of influenza virus. Proc Natl Acad Sci U S A 87:3802–3805

Schnell MJ, Mebatsion T, Conzelmann KK (1994) Infectious rabies viruses from cloned cDNA. EMBO J 13:4195–4203

Preface

Genome nucleotide sequencing was completed for the representative members in the order *Mononegavirales* by the end of the 1980s. The next challenge in this research field was to establish a system to recover, or rescue, an infectious virus entirely from cDNA and hence allow reverse genetics. The competition to be the first in establishing this feat took years because it required a complex procedure with delicate, fine tuning. In 1994, at the end of a decade of trial-and-error experimentation, Conzelmann and his colleagues eventually succeeded in the rescue of rabies virus in the *Rhabdoviridae* from a cDNA plasmid. Shortly thereafter, six research groups achieved rescue from the respective cDNAs of Sendai virus (SeV), measles virus, and respiratory syncytial virus in the family *Paramyxoviridae* and another rhabdovirus, vesicular stomatitis virus. By the end of the 1990s, the technology became available for most species including newly emergent deadly agents such as Ebola virus and Nipah virus in the *Mononegavirales*. It has become possible to manipulate their genomes at will and evaluate the outcomes not only in terms of viral proliferation at the cellular level but also in the context of viral multiplication and pathogenesis in the entire host organisms, settling outstanding issues and resolving enigmas in the biology of *Mononegavirales*.

For certain virus species, reverse genetics has evolved into technology to use these viruses as an expression vector. This type of vector is regarded as a novel class, cytoplasmic RNA vector, because these viruses fulfill their life cycle entirely in the cytoplasm without having a nuclear phase, raising theoretically no fear of genotoxicity including integration into host chromosomes. Studies to develop the SeV vector appear to typify this line of research because they have demonstrated great potential for diverse applications and because some of them have now left or are about to leave the laboratory bench for preliminary clinical trials. Such progress in SeV technology appears to be largely the result of some characteristics intrinsic to this creature, including its nonintegrating nature, extremely high production titers, extremely high performance of *trans*-gene expression, remarkable target cell breadth, and no potential pathogenicity for humans. While active incorporation of the basic knowledge of SeV biology has accelerated the production of target-oriented devices to be used in special settings, the preclinical and clinical research outcomes in turn stimulate further research at the basic level, illustrating that science and technology spiral up, interacting synergistically with each other.

This book starts with the afore-described landmark work on rabies virus that opened the era of *Mononegavirales* reverse genetics and its exciting extension, and then incorporates SeV-specific chapters including its biology and reverse genetics, the concept and technology underlying SeV vector development for use in human medicine, and actual applications in five different special areas.

The journey, starting here, to the ultimate goal will be hard and long. Many of the authors of this book are relatively young, not established authorities. I hope that their continued efforts will eventually bear fruit. Finally, I express my cordial appreciation to all of the authors. I would also like to express my deepest gratitude to Dr. Peter Palese for having contributed the insightful and encouraging Foreword.

Tokyo, Japan
31 January 2014

Yoshiyuki Nagai

Contents

Chapter 1
Reverse Genetics of *Mononegavirales*: The Rabies Virus Paradigm

Karl-Klaus Conzelmann

Abstract The neurotropic rabies virus (RABV) is a prototype member of the *Mononegavirales* order of viruses and is the most significant human pathogen of the *Rhabdoviridae* family. A reverse genetics system for RABV was established almost 20 years ago, providing a paradigm for other *Mononegavirales* members as well. The availability of engineered recombinant viruses opened a new era to study common aspects of *Mononegavirales* biology and specific aspects of the unique lifestyle and pathogenesis of individual members. Above all, the knowledge gained has allowed engineering of beneficial biomedical tools such as viral vectors, vaccines, and tracers. In this chapter, the development of the classical rabies virus reverse genetics approach is described, and some of the most exciting biomedical applications for recombinant RABV and other *Mononegavirales* are briefly addressed.

1.1 *Mononegavirales*: A Huge Diversity of Similar Viruses

The order *Mononegavirales*, or nonsegmented negative strand RNA viruses (NNSV), comprises the families of *Rhabdoviridae*, *Paramyxoviridae*, *Filoviridae*, and *Bornaviridae*. Members of the *Rhabdoviridae* family have the broadest host range among the *Mononegavirales* and infect plants, insects, fish, aquatic, aerial, and terrestrial animals, and humans (for reviews, see Fu 2005; Pringle 2005). Remarkably, the only globally important human pathogen among rhabdoviruses is the rabies virus (RABV) of the genus *Lyssavirus*, which causes rabies encephalitis, a long-known, most dangerous and feared zoonotic disease. In spite of the availability of potent vaccines for animals and humans and effective postexposure

K.-K. Conzelmann (✉)
Max von Pettenkofer-Institute and Gene Center, Ludwig-Maximilians-University Munich, Feodor-Lynen-Str. 25, 81377 Muenchen, Germany
e-mail: conzelma@lmb.uni-muenchen.de

Y. Nagai (ed.), *Sendai Virus Vector: Advantages and Applications*,
DOI 10.1007/978-4-431-54556-9_1, © Springer Japan 2013

treatments, rabies still causes more than 50,000 human deaths per year in rural areas of Asia and Africa, where it is mostly transmitted by rabid dogs. RABV is a typical member of the *Mononegavirales* in terms of virus organization and gene expression, and in this respect is closely related to the insect-transmitted vesicular stomatitis virus (VSV), a well-studied prototype rhabdovirus of the *Vesiculovirus* genus. However, the biology of RABV is unique in terms of direct transmission between mammals, a strict neurotropism, and an extremely broad host species range. These features are key to rabies biology, transmission, and pathogenesis, and represent severe obstacles in terms of RABV eradication.

In contrast to the *Rhabdoviridae*, numerous members of the *Paramyxoviridae* family have evolved to become human pathogens, such as the measles, mumps, parainfluenza, respiratory syncytial, Hendra, and Nipah viruses. An extremely valuable model paramyxovirus has always been Sendai virus (murine parainfluenza virus type 1; hemagglutinating virus of Japan, HVJ), the study of which has uncovered numerous traits of paramyxovirus biology (Nagai et al. 2011), and the use of which as a biomedical tool is highly promising (see other chapters in this volume). Filoviruses such as Ebola and Marburg virus and Bornaviruses are animal viruses, which have not (yet) managed to become established in the human population (Lamb 2007).

As different as individual members, genera, or families of *Mononegavirales* may be in terms of shape, biology, and pathogenesis, they all are but variations of a common theme of viral genome and particle organization and mode of gene expression. The single 10- to 20-kb RNA genomes and antigenomes of *Mononegavirales* exist in the form of a permanent helical ribonucleoprotein (RNP), in which the RNA is enclosed in a nucleoprotein (N, or NP). The information encoded is expressed by sequential and polar transcription of discrete subgenomic mRNAs from the RNPs (Fig. 1.1) (for detailed review, see Whelan et al. 2004). The polymerase seems to be an RNP-dependent RNP polymerase for replication of RNPs and an RNP-dependent RNA polymerase for transcription of subgenomic mRNAs. It is composed of a large (L) multifunctional catalytic protein and a noncatalytic cofactor, mostly named phosphoprotein (P). The existence of such P protein distinguishes *Mononegavirales* from related segmented NSV such as bunyaviruses and arenaviruses. The most highly conserved N and L proteins of the *Mononegavirales* families still share common sequence blocks, whereas P proteins are much more variable. In addition to N, P, and L, *Mononegavirales* encode at least one matrix protein (M) and at least one transmembrane glycoprotein (e.g., G, GP, F), which make up the viral envelope, and which are predominantly engaged in formation of RNP-containing virions and infection of new target cells, respectively. Notably, the order of these five minimal genes (3′-N-P-M-G-L-5′) is strictly conserved in all *Mononegavirales* genomes (Fig. 1.1). This conservation is thought to reflect the stoichiometric need of the individual proteins, which is brought about through sequential and polar transcription. The modular organization of *Mononegavirales* genomes facilitates the acquisition of additional genes, mostly to internal positions, to specifically adapt to novel hosts and environments. In addition, individual genes of *Mononegavirales* may encode multiple gene products, as illustrated by the accessory proteins expressed from the paramyxovirus "P" gene (Sakaguchi et al. 2008). The versatility of the

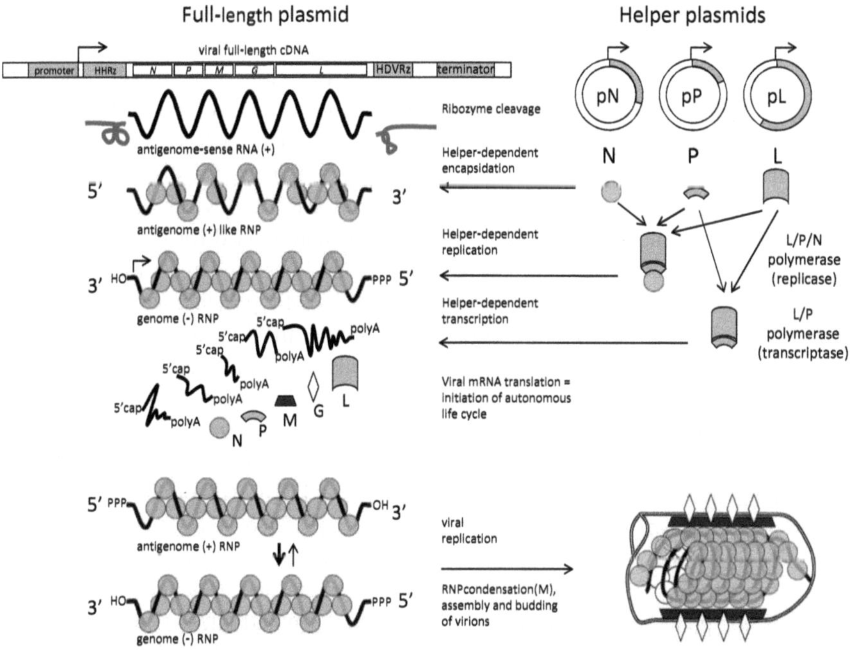

Fig. 1.1 Recovery of *Mononegavirales* from plasmids. The standard protocol for recovery of *Mononegavirales* from cDNA involves simultaneous transfection of at least four plasmids into a cell, one comprising the full-length cDNA of the virus and three expression plasmids encoding helper virus proteins N, P, and L. The viral cDNA is oriented to give rise to a positive strand (antigenomic) RNA and is flanked by autocatalytic ribozyme sequences to produce the correct viral 5′- and 3′-ends upon transcription of the RNA by T7 RNA polymerase (encoded in transgenic cells, or from transfected plasmids) or by an endogenous cellular polymerase such as polymerase II. The virus-like RNA must be encapsulated into N protein to mimic a ribonucleoprotein (RNP) and to be recognized by the viral polymerase, which can act as a replicase (L–P–N complex) producing complementary RNPs, or as a transcriptase (L–P complex), producing nonencapsidated subgenomic mRNAs. Because the viral N, P, M, G, and L mRNAs are produced from the genome RNP, an autonomous infection cycle is initiated. Viral replication gives rise to an excess of genome RNPs, which can be used for assembly of novel virions

common NNSV blueprint is illustrated by the fact that *Mononegavirales* have conquered different kingdoms of life, including plants and animals, as well as numerous niches within a single species.

1.2 Synthetic Biology or "Reverse Genetics" of Viruses

Viruses and phages can be regarded as highly mobile genetic elements that need only a cell to translate the encoded information into protein. In the past, they have played a major role both as research objects and as tools in the development of modern gene technology. They are also at the forefront of synthetic biology. In fact, introducing authentic or modified nucleic acids or cDNA of viruses into a cell in many cases leads to initiation of an infectious cycle and ultimately in the creation of viruses never existing previously in nature. This technique was initially found to work for DNA viruses such as SV40 (Goff and Berg 1976), retroviruses (Rothenberg et al. 1977), and RNA phages (Taniguchi et al. 1978), and for RNA viruses such as poliovirus (Racaniello and Baltimore 1981), now known as "positive" strand RNA viruses. Importantly, the RNA of the positive strand RNA viruses, or the RNA derived from the introduced cDNA, can be immediately used for translation of all virus proteins. The negative strand RNA virus particles, however, were found to comprise RNA complementary to mRNA, and a specific polymerase (which is not present in cells) to transcribe individual subgenomic mRNAs upon entry into cells (Baltimore et al. 1970; Baltimore 1971). Obviously, the cDNA or cDNA-derived RNAs of *Mononegavirales* are not infectious because they are not suitable for translation of the necessary viral proteins by the cellular machinery. It was therefore noted by Racaniello and Baltimore (Racaniello and Baltimore 1981) that "It is uncertain whether the application of recombinant DNA techniques to negative-strand RNA viruses (5) [referring to Baltimore 1971] or viruses with multiple genome RNA's will also yield infectious cDNA." The term "reversed genetics" for this approach was introduced by Charles Weissmann and colleagues, describing methods for generating point mutations at predetermined sites of RNA or DNA genomes, specifically with Qbeta phage RNA as a template (Weissmann et al. 1979). Fortunately, in spite of inherent drawbacks, reversed genetics has become available for negative strand RNA viruses as well, including members of all *Mononegavirales* virus families, the *Rhabdoviridae*, *Paramyxoviridae*, *Filoviridae*, and *Bornaviridae*.

1.3 Recovery of *Mononegavirales* from cDNA

1.3.1 Rescue and Use of Model Minigenomes

Successful rescue of viruses from recombinant RNA relies on the expression of viral proteins from the viral RNA. In NNSV, expression of the encoded proteins is achieved by sequential transcription of individual mRNAs from the negative strand RNP. The proposed stop–restart model suggests entry of the polymerase exclusively at the 3′-end of the genomic (negative strand) RNP and the sequential transcription of a short nonmodified leader RNA specified by the 3′-end of the genome and of 5′-capped and 3′-polyadenylated mRNAs. The ends of the subgenomic RNAs are specified by conserved transcription stop/polyadenylation and restart signals at the gene borders. New RNPs are only formed upon accumulation of sufficient N protein, which apparently switches the transcriptase form of the polymerase to a replicase form of the polymerase. In this "replication" mode of the polymerase, synthesis of RNA and encapsidation into N protein is mechanistically linked, and the transcription signals are ignored.

The expression of viral proteins such as the polymerase components L and P is straightforward by using so-called "helper" plasmids or virus vectors providing the proteins in *trans* (Fig. 1.1). In contrast, the encapsidation of viral RNA by N protein, that is, the reconstitution of a functional RNP from premade RNA and N protein, also called "illegitimate encapsidation," turned out to be more difficult. Such artificial RNP must be able to serve as a template for the polymerase. Early experimental work mostly involved short defective interfering (DI) RNAs, or minigenomes comprising only the noncoding terminal promoter regions of *Mononegavirales* genomes. Initial encouraging success was obtained with the segmented influenza virus by transfection of an *in vitro* packaged genome segment (Luytjes et al. 1989), and for the *Mononegavirales* with Sendai virus (Park et al. 1991) by transfection of an *in vitro*-transcribed minigenome RNA into Sendai helper virus-infected cells, which resulted in expression of the minigenome-encoded reporter gene.

A major breakthrough was then the establishment of helper virus-free systems, in which DI- or minigenome RNAs and virus proteins were simultaneously expressed within cells from transfected circular plasmids (Pattnaik et al. 1992; Pattnaik and Wertz 1990, 1991). The system employed vaccinia virus (vv)-encoded phage T7 RNA polymerase and plasmids equipped with T7 RNA polymerase promoter and terminator sequences (Fuerst et al. 1986). The critical ends of the model genome RNAs were generated by self-cleaving ribozymes (Fig. 1.1), such as the antigenome ribozyme of hepatitis delta virus (HDVagrz) (Perrotta and Been 1990; Sharmeen et al. 1988). The vv/T7 protein and RNA expression system paved the way for similar progress with minus strand RNA minigenomes of all families of the *Mononegavirales*. This achievement provided immediate experimental access to the clarification of fundamental and longstanding questions in the *Mononegavirales* field, including the identity and function of *cis*-acting signal sequences such as terminal promoters and transcription signals, or other RNA-based regulatory mechanisms, just to mention the rule of six (Calain and Roux 1993; Vulliemoz and Roux 2001), as well as the contribution of individual proteins to gene expression and replication. Moreover, minigenome systems are valuable tools for high-throughput screening of antivirals and provide a safe possibility to study aspects of highly dangerous viruses (Biacchesi 2011; Conzelmann 2004; Marriott and Easton 1999; Theriault et al. 2005; Whelan et al. 2004).

1.3.2 Rescue of Recombinant Viruses from cDNA

Although minigenome recovery in the described system worked reliably for many virus species, the desired generation of recombinant nondeficient viruses using the same approaches seemed to be out of reach for some time. Finally, an approach in which positive sense antigenome RNA rather than genome sense RNA of RABV was employed for initial encapsidation resulted in the successful recovery of the first negative strand RNA virus entirely from cDNA in our laboratory (Schnell et al. 1994). In fact, this seemingly counterintuitive "positive approach" circumvents lethal common problems associated with genome sense RNA expression and encapsidation and was immediately applicable for recovery of other *Mononegavirales* as well, including rinderpest virus, respiratory syncytial virus, parainfluenza virus type 3, Sendai virus, SV5 or parainfluenza virus type 5, measles virus, VSV (Baron and Barrett 1997; Collins et al. 1995; Durbin et al. 1997; Garcin et al. 1995; He et al. 1997; Hoffman and Banerjee 1997; Kato et al. 1996; Lawson et al. 1995; Radecke et al. 1995; Whelan et al. 1995), and even of segmented NSV such as Bunyamwera virus (Bridgen and Elliott 1996). Although initial encapsidation of antigenome RNA requires that an extra step of replication is supported by the coexpressed N, P, and L helper proteins (see Fig. 1.1), it avoids deleterious hybridization of viral genome RNA and the complementary helper protein mRNAs expressed simultaneously in the same cell (Conzelmann 1996; Roberts and Rose 1998; Schnell et al. 1994). The resulting dsRNA is assumed not only to interfere with RNA encapsidation and helper protein translation but also to trigger innate antiviral immunity (Randall and Goodbourn 2008; Rieder and Conzelmann 2011). Although most laboratories confirmed a failure of *Mononegavirales* genome sense RNA rescue, in two highly optimized systems viruses could be rescued from both genome and antigenome RNAs, namely Sendai virus (Kato et al. 1996) and HPIV-3 (Durbin et al. 1997). In the former work, the magnitude of the antisense problem was nicely illustrated, as rescue efficiency with Sendai virus negative strand RNA was at least 100 fold less effective than with positive strand RNA (Kato et al. 1996).

1.3.3 Technical Improvement of Virus Rescue Systems

The principle of the "classical" vaccinia virus/T7 virus rescue system turned out to work for viruses of all *Mononegavirales* families, including nonmammalian species such as fish rhabdoviruses (Biacchesi 2011). Variations mostly included the use of other sources of T7 RNA polymerase, such as the host range-restricted vaccinia virus MVA-T7, or other poxvirus species, alleviating the need for active removal of the T7-encoding helper viruses just by passaging the recoveries in nonpermissive cells (Conzelmann 2004). In addition, helper virus-free systems have been developed, based on transient expression of T7 RNA polymerase from plasmids or in stable cell lines. Particularly, BSR T7/5 cells (Buchholz et al. 1999) are being widely used, as this cell clone combines the advantage of high-level T7 RNA polymerase expression, and of having a defect in the IRF3 gene, such that the antiviral interferon response is not induced (Conzelmann 2004). In the absence of vaccinia virus capping and polyadenylation enzymes, T7 RNA polymerase transcripts have 5′-triphosphate ends, which is advantageous for generating virus-like antigenomes, but detrimental for helper protein translation. Therefore, T7 protein expression constructs should include an upstream IRES element, or helper proteins should be expressed from polymerase II promoter-driven plasmids.

Of particular importance for rescue of *Mononegavirales* are the ends of the RNA to be packaged, as already noticed in many minigenome systems. Particularly, a precise 3′-end was found critical for initial recognition by the viral polymerase and replication, whereas some extra nonviral nucleotides at the 5′-end were well tolerated (Pattnaik et al. 1992). Therefore, in most *Mononegavirales* rescue systems T7 promoter transcripts have been employed that comprise three extra G residues to facilitate transcription initiation by T7 RNA polymerase (Conzelmann and Schnell 1994; Pattnaik et al. 1992). However, precise 5′-ends of the transcripts do more than compensate for lower transcript levels, as illustrated by highly efficient rescue of Sendai virus RNA lacking extra G residues (Kato et al. 1996). More recently, hammerhead ribozymes (Blount and Uhlenbeck 2002) are being increasingly employed for generation of exact 5′-ends of transcripts (Fig. 1.1). A direct comparison of RABV strain SAD L16 cDNA constructs yielding transcripts comprising three extra G residues or possessing correct 5′-ends generated by a hammerhead ribozyme revealed a tenfold advantage of the latter (Ghanem et al. 2012). In this work it was also noticed that the cleavage efficiency of the HDV agrz used so far in almost all rescue systems is very low in transfected cells, and that only 10 % of the RNA has the right 3′-end available for initiation of replication. Exchange with a longer and more effective HDV ribozyme (SC) cleaving 90 % of intracellular transcripts again increased rescue efficiency more than tenfold. Moreover, the combination of hammerhead and SC HDVagrz in full-length clones of the RV SAD L16 (pSAD HH-L16-SC) had a synergistic effect and improved rescue by 100 fold, yielding the most efficient RABV rescue plasmid so far (Ghanem et al. 2012).

The use of polymerases such as T7 RNA polymerase has the advantage of allowing transcription of transfected plasmids in the cytoplasm and therefore is ideal for the generation of viruses with a cytoplasmic replication cycle. Specifically, potential problems with nuclear export of RNA transcripts synthesized by cellular polymerases such as RNA Pol II are circumvented. Nevertheless, CMV promoter-driven rescue has been achieved for several cytoplasmic mammalian and fish rhabdoviruses (Ammayappan et al. 2010; Biacchesi 2011; Huang et al. 2010; Inoue et al. 2003; Ming et al. 2009; Orbanz and Finke 2010; Tao et al. 2010), as well as paramyxoviruses (Martin et al. 2006), suggesting the possibility to conditionally express viruses from host cell genome-encoded DNA in cultured cells or in animals.

1.4 Genetic Engineering of *Mononegavirales* Genomes

Even the highly optimized *Mononegavirales* rescue systems remain less effective by orders of magnitude than positive strand RNA rescue. The major bottleneck is most probably a very poor rate of illegitimate RNA encapsidation by the N protein. Although this precludes a few applications, such as the use of *Mononegavirales* as cloning vectors for genetic libraries, it is sufficient for rescue of individual viruses, even severely attenuated viruses or gene-defective viruses, which require additional complementation in *trans*. Nonetheless, the possibilities of genetic manipulation of individual *Mononegavirales* as well as the applications are enormous as a result of their particular genome organization, lack of packaging size, the stability of their RNA genomes, and the lack of recombination with host sequences.

1.4.1 Mutation of Virus Functions: Host Immune Escape

In the past two decades the possibility of site-directed mutagenesis of *Mononegavirales* has brought about hundreds of publications and a wealth of information on the structure, function, biology, virus–host interplay, and pathogenesis of individual *Mononegavirales.* One recent common topic is the interplay of viruses with the host innate immune system. This field has been fueled by tremendous progress in the identification of pattern recognition receptors (PRR) and details of the signaling pathways leading to the expression of type I and III interferons (IFN) and of proinflammatory cytokines.

IFN and proinflammatory cytokines are activated by two PRR families recognizing non-self viral RNA, the endosomal transmembrane Toll-like receptors (TLR 3, 7/8) and the cytoplasmic (RIG-I)-like helicases RIG-I (retinoic acid inducible gene-I, also known as DDX58) and MDA5 (melanoma differentiation-associated gene 5, also known as IFIH1 or helicard) (Kato et al. 2011; Kawasaki et al. 2011). The pathways for IFN induction merge in the activation of interferon regulatory factors (IRF) 3 and IRF7, which control transcription of type I and III IFN genes (Honda et al. 2006; Onoguchi et al. 2007; Yoneyama et al. 1998). The pathways for induction of proinflammatory cytokines such as tumor necrosis factor (TNF)-α and interleukins lead to canonical activation of NF-κB, which also supports transcription of the early IFNs (IFN-β, IFN-α4).

It turns out that *Mononegavirales* infection of cells is generally and effectively sensed by the ubiquitous RIG-I, which recognizes viral triphosphate RNAs (Cui et al. 2008; Hornung et al. 2006), that is, the leader RNA and possibly non-encapsidated genome RNAs of *Mononegavirales* (Gerlier and Lyles 2011). Specific dsRNA patterns recognized by MDA5 remain elusive, and the usually minor contribution of MDA5 may differ for both virus species and host cell types (Bruns and Horvath 2012; Loo and Gale 2011). RIG-I or MDA5 activation by viral RNAs leads to their association with the mitochondrial protein IPS-1 (also known as MAVS, VISA, or Cardif), via CARD domains, and to the recruitment of a signaling complex in which IRF3 is phosphorylated by the kinases TBK-1 and IKKi (also known as IKKε) (Fitzgerald et al. 2003; Sharma et al. 2003). Ser386-phosphorylated IRF3 dimerizes and is imported into the nucleus where it drives transcription of IFN-β mRNA in conjunction with NF-κB and AP1 transcription factors (Ford and Thanos 2010). The secreted IFN acts on cells by activating JAK/STAT signaling pathways (Platanias 2005), which activate hundreds of IFN-stimulated genes (ISG), many of which have antiviral and immunomodulatory activities (Schoggins et al. 2011; Theofilopoulos et al. 2005).

Notably, even viruses with a very small coding capacity, such as the *Mononegavirales*, have evolved means to interfere with both IFN induction and IFN-mediated response and with ISG function, illustrating the power of the IFN system and the high evolutionary pressure for viruses to evolve suitable countermeasures (Gerlier and Lyles 2011; Goodbourn and Randall 2009; Randall and Goodbourn 2008; Versteeg and Garcia-Sastre 2010). Proper encapsidation of the viral RNA by N proteins into RNPs may be a common advantageous trait of

Mononegavirales and largely prevent recognition by PRRs, but this requires a well-balanced synthesis of RNA and protein. If this delicate balance is disturbed, possibly in cell types not supporting RNA and protein synthesis equally well, or by overproduced DI RNAs, virus infection is easily recognized. In addition to shielding viral RNA, *Mononegavirales* have developed means to actively interfere with signal transduction. In mammalian *Mononegavirales*, these IFN antagonists are typically encoded by the P genes.

The RABV P protein, in addition to its functions as an N chaperone and polymerase cofactor, was found to have important roles in counteracting multiple specific steps of IFN gene expression, IFN-induced STAT signaling, and the function of antiviral proteins (Rieder and Conzelmann 2011). Specifically, RABV P targets the activation of the transcription factors IRF3 and IRF7 and prevents their phosphorylation by the kinases TBK-1 and IKKi. The IRFs are therefore not able to dimerize, and their import into the nucleus and transcription of the IFN genes is precluded (Brzózka et al. 2005). In addition, STAT-mediated transcription of ISGs is precluded in the presence of RABV P. After binding of IFN to the IFN receptor, STAT1 and STAT2 are phosphorylated at specific tyrosine residues by receptor-associated Janus kinases, which is a prerequisite for hetero-dimerization, association with IRF9, and nuclear import of the STATs (Platanias 2005). Most remarkably, RABV P binds exclusively to tyrosine-phosphorylated STAT1 and STAT2, whereas in nonactivated cells an association of P with any STAT is not apparent (Brzózka et al. 2006). Targeting of an already activated form of STAT is unprecedented and may appear critical in terms of timing. However, such conditional emergency activity may allow P to perform its many other functions in virus replication. The high relevance in the virus context of both IFN antagonistic functions was recently illustrated by generating recombinant viruses expressing low levels of P (Brzózka et al. 2005), P mutants lacking specifically the IRF inhibitory function (Rieder et al. 2011), or viruses defective in STAT inhibition (Ito et al. 2010).

The P genes of paramyxoviruses such as Sendai virus encode, in addition to P, "accessory" proteins such as V or C, which largely take over IFN escape functions (Nagai et al. 2011; Sakaguchi et al. 2008) (see Chap. 2 for details). The V proteins of paramyxoviruses are translated from an edited P mRNA and possess a specific short C-terminal domain (V-CTD) of conserved structure and which is involved in binding and inhibiting a variety of target proteins involved in IFN induction (MDA5, IKKα, and IRF7), IFN signaling (STAT1, STAT2, Janus kinases), and NF-κB pathways (p65/RelA) (Gerlier and Lyles 2011; Goodbourn and Randall 2009; Motz et al. 2013; Schuhmann et al. 2011). The C proteins, which are expressed from an alternative reading frame of the P gene, seem to be involved in actively counteracting IFN (Sparrer et al. 2012), but the C proteins of some viruses such as the measles virus may also be active in controlling viral RNA synthesis and preventing dsRNA accumulation, thereby limiting IFN responses (Boonyaratanakornkit et al. 2011; Pfaller et al. 2013). Notably, V and C proteins from different paramyxovirus species and genera may differ in their affinity to individual targets, probably reflecting adaptation to counteract pathways important in specific hosts, organs, and cell types (Versteeg and Garcia-Sastre 2010).

In striking contrast to RABV and paramyxoviruses, innate immune escape of the insect-transmitted rhabdovirus VSV seems not to be determined by P gene functions. In VSV, rather, the M protein seems to represent the major factor, by shutting down host gene transcription and mRNA export (Rajani et al. 2012 and references therein).

The study of the structure of immune-stimulating RNAs of *Mononegavirales* and of the viral inhibitory mechanisms is not only shedding light on the nature of the innate immune system but offers directions toward the development of better attenuated and more immunogenic live vaccines. The elimination or modification of viral IFN antagonists by reverse genetics is yielding viruses that activate both innate and adaptive immune responses better than the wild-type (wt) virus.

1.4.2 Reorganization of Genomes: Vaccines, Vectors, Tracers, and Other Tools

The modular organization of genes in the *Mononegavirales* genomes (Fig. 1.1) allows easy deletion or insertion of extra genes without disturbing the expression machinery or virus formation. Gene cassettes comprising the transcriptional signals for transcription start and stop can be inserted between any of the viral genes, and in some viruses even in the 3′- and 5′-terminal positions. The major effect of extra transcription units in most viruses seems to be restricted to a modest transcriptional attenuation of the downstream genes. Thus, for example, up to four extra genes have been introduced successfully into the genomes of Sendai virus to simultaneously express multiple transcription factors for cell reprogramming (Nishimura et al. 2011). Transcriptional attenuation can be compensated by equipping the downstream genes with gene borders known to cause less attenuation than others (Finke et al. 2000). Alternative approaches include construction of single multi-cistronic transcription units encoding multiple proteins, by using picornaviral IRES elements or 2A-like sequences (Marschalek et al. 2009). By manipulation of the genome and antigenome promoters, even rabies and Sendai viruses with an ambisense gene expression strategy have been recovered. These viruses express the set of viral proteins from the viral genome RNA and extra proteins from the viral antigenome RNA (Finke and Conzelmann 1997; Le Mercier et al. 2002).

As the position in the viral genome determines the relative expression levels, shifting of genes to different positions is a versatile and commonly used tool to study dose effects of both viral and foreign genes. In combination with other approaches such as gene deletion, the gene shift approach has been used in the case of RABV to identify so far unappreciated roles of M and P proteins as a regulator of viral transcription (Finke et al. 2003) or as an interferon antagonist (Brzózka et al. 2005), respectively.

1.4.2.1 Vectors for Vaccines and Therapeutic Genes

The ability to express foreign or heterologous genes and antigens is the basis for the development of heterologous or multivalent vaccines. Particularly, many

Mononegavirales are well known to induce strong humoral, cellular, and mucosal immune responses. Established human vaccine virus strains, such as the measles Schwarz strain of the Edmonston lineage, or animal viruses that do not cause disease in humans such as Sendai virus or Newcastle disease virus, seem to be immediately suited as promising carriers. In fact, numerous viral antigens from the major agents threatening human health, including human immunodeficiency virus (HIV)-1, Ebola virus, influenza virus, and SARS virus, hepatitis C virus (HCV), and HBV, have been expressed from these viruses (Brandler et al. 2010; Geisbert and Feldmann 2011; Khattar et al. 2011; Yu et al. 2010), and the first human trials are awaited. Also, recombinant RABV is being exploited as a vector vaccine, particularly for immunization against HIV-1 (Gomme et al. 2011), because completely attenuated RABV variants are now available. The RABV G protein is the major virulence factor of RABV and is responsible for the pronounced neurotropism and neuroinvasiveness of the virus. Mutations affecting the arginine 333 residue of the G protein dramatically alter the cell tropism and render RABV avirulent, even after intracerebral injection (Mebatsion 2001). Similarly, RABV lacking the IRF3 inhibitory function are completely attenuated after infection in mouse brains (Rieder et al. 2011).

1.4.2.2 Envelope Switching

A particular advantage of *Mononegavirales* is their amenability to envelope switching. As confirmed early by gene deletion mutants, the G proteins of the rhabdoviruses RABV and VSV are not essential for virus formation and budding of virions, although they may contribute to the efficiency of the process (Mebatsion et al. 1996; Schnell et al. 1998). Moreover, they can be entirely replaced by foreign type I transmembrane proteins. By using viral proteins active in receptor binding and membrane fusion, virions with a novel tropism or host range can be generated (Johnson et al. 1997; Mebatsion and Conzelmann 1996; Schnell et al. 1996). Even viruses specifically targeting cells infected with another virus can be generated (Mebatsion et al. 1997; Schnell et al. 1997). Natural paramyxoviruses have been long known for phenotypic mixing, or formation of pseudo-type viruses in coinfections (Kimura 1973), and are similarly amenable to artificial envelope swapping (for a recent paper, see Mourez et al. 2011). However, the requirements for incorporation of heterologous envelope proteins may differ for viruses. In contrast to VSV, which readily incorporates numerous type I transmembrane proteins (Schnell et al. 1998), efficient incorporation into RABV requires a C-tail sequence and structure similar to that of RABV G (Mebatsion and Conzelmann 1996).

Envelope swapping is a strategy extremely useful for various approaches. For vaccination purposes, sequential application of viruses possessing different envelopes avoids immune recognition and allows repeated use of the same vector backbone. In addition, the glycoproteins of highly dangerous viruses can be studied safely by using pseudo-type virions. Appropriate de- and retargeting of viruses is especially important in the field of oncolytic virotherapy. Envelope switching can involve the generation of chimeric "surrogate" viruses, in which a novel G-protein gene replaces the autologous gene, or of pseudo-type viruses, where the G gene is deleted and the protein is provided in *trans*. The latter is a safe approach, as single-round vectors are being generated that can infect a single cell but cannot spread further.

1.4.2.3 Oncolytic Virotherapy

A highly promising field for some recombinant *Mononegavirales* is oncolytic virotherapy, which means the use of viruses to selectively infect and damage cancerous tissues without causing harm to normal tissues (Russell et al. 2012). Preferred replication in transformed cells is partly caused by their genetic defects in innate immune responses, such as impaired STAT signaling or induction of apoptosis, that is, in major antiviral mechanisms. Viruses with defects in their IFN antagonists (such as paramyxovirus V proteins, or VSV M protein) therefore may replicate effectively in such cancer cells, whereas they are attenuated in normal tissue in which the antiviral mechanisms are intact. Some viruses, such as Newcastle disease virus and mumps virus, seem to have a strong natural preference for cancer cells, but other viruses such as measles, VSV, or Sendai virus can be readily engineered to make them more cancer specific (Cattaneo et al. 2008; Kinoh et al. 2009). Engineering of oncolytic viruses may include retargeting by destroying the natural receptor-binding domains of the viral envelope proteins in combination with displaying polypeptide ligands on the surface to facilitate infection of tumor cells overexpressing the targeted receptor. An outstanding series of retargeted measles viruses has been created accordingly, targeting various antigens overexpressed on tumor cells (Cattaneo et al. 2008). An alternative approach is complete envelope switching, that is, the complete exchange of the viral surface proteins. Promising examples of envelope switching include the use of oncolytic VSV carrying the envelope of LCMV, enhancing the infectivity for glioma cells and minimizing neurotropism (Muik et al. 2011). In the case of Sendai virus, tumor specificity could be enhanced by exploiting the need for proteolytic activation of the fusion protein (F). Specifically, the F-cleavage site was modified in such a way that it was cleaved by a protease highly expressed only in tumor cells (Morodomi et al. 2012).

1.4.2.4 Monosynaptic Tracing of Neurons with RABV: Envelope Swapping in Situ

The large group of *Mononegavirales* comprises viruses with highly diverse biological traits and thus represents an almost unlimited source of tools for all kinds of basic research and biomedical applications. In the past two decades of *Mononegavirales* reverse genetics, Sendai virus has emerged as a prime example for the broad and huge potential of this virus, as impressively illustrated in detail in the following chapters. Major inherent advantages of Sendai virus include its apathogenicity for humans and broad host and tissue range. Also, however, virulent viruses and, more explicitly, specific pathogenic traits of these viruses can be exploited for research. One example is the unparalleled neurotropism of RABV, which makes this virus a fierce pathogen but also offers unique possibilities for the study of the nervous system. RABV is entirely adapted to the nervous system of its hosts and—importantly—is completely dependent on the integrity of the peripheral nervous system to reach the brain where it replicates best and where it can elicit behavioral changes of the host to facilitate virus transmission within the population. Upon peripheral infection, RABV is taken up by the axon ends of neurons via endocytosis and is retrogradely transported in axonal transport vesicles toward the cell body (Klingen et al. 2008). Upon membrane fusion and release of the RNP into the cytoplasm, replication takes place without severe effects on the viability and function of neurons (Lafon 2011). Remarkably, newly assembled virions seem to be released exclusively at functional synapses where they can enter second-order neurons at the presynaptic membrane (Astic et al. 1993; Ugolini 1995). For trans-synaptic transmission, the RABV G protein is required (Etessami et al. 2000).

This exclusive trans-synaptic spread is unique among viruses and therefore natural RABV has been used for years as a "polysynaptic" tracer in specialized laboratories (Dum and Strick 2012). Direct synaptic connections of individual neurons, however, cannot be easily determined with conventional nonviral or viral tracers. The availability of RABV reverse genetics and progress in vector construction and virus retargeting, however, has more recently allowed the development of the first system for "monosynaptic" tracing of direct neuronal connections (Wickersham et al. 2007) (Fig. 1.2).

The system involves targeted infection of a defined (postsynaptic) neuron with a G gene-deficient recombinant RABV (ΔG RABV) expressing a fluorescent protein such as green fluorescent protein (GFP). Specific targeting of the postsynaptic neuron is achieved, for example, by expression of the avian TVA receptor, and infection with ΔG RABV pseudo-typed with EnvA of an avian retrovirus that uses the TVA receptor for entry (Fig. 1.2). Additional expression of RABV G protein supports trans-synaptic transfer of the virus to synaptically connected (presynaptic) neurons. As no G is expressed in the presynaptic cells, the virus does not spread further. This safe ΔG RABV system is now being used widely by neurobiologists not only to dissect sensory and motor circuits in the nervous system of various animal models but also to read out and modulate the activity of single neurons and of neuronal circuits (Ginger et al. 2013; Wickersham and Feinberg 2012). It is expected that recombinant RABV in this way greatly contributes to the understanding of how our most complex brains work.

1.5 Conclusions

After an intricate birth phase, reverse genetics of *Mononegavirales* has become a standard technique to study this group of medically important viruses. Tremendous progress has been made subsequently in understanding the biology of these viruses,

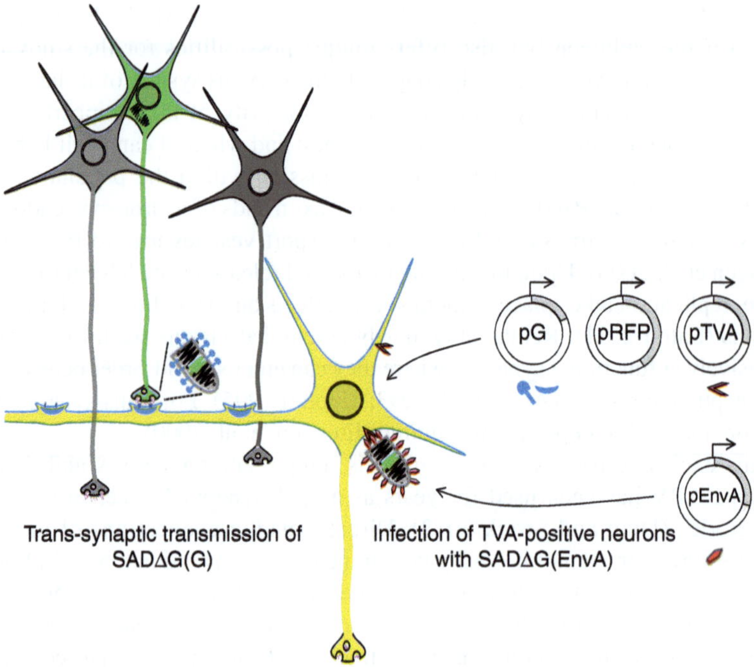

Fig. 1.2 Monosynaptic tracing of neuronal circuits with rabies virus (RABV) ΔG. The exclusive trans-synaptic transmission of RABV is being exploited for mapping of direct connections between neurons. A starter neuron (*yellow*) that provides RABV G protein (*blue*) is transmitting a G gene-deleted RABV expressing eGFP (SAD ΔG-eGFP) via a synapse to presynaptic neurons (*green*). Because the G protein is not produced in the presynaptic neurons, the virus cannot be transmitted further. Selective infection of the postsynaptic starter neuron can be achieved, for example, by expression of a specific receptor (here, TVA), making cells permissive for infection with SAD ΔG-eGFP pseudo-typed with a retroviral envelope protein (EnvA)

the interplay with their hosts, and the mechanisms of pathogenicity. It is becoming increasingly clear that the host innate immune system and specific viral counteractions profoundly shape virus–host relationships. Reverse genetics can now be used to make friends from foes, by converting the viruses to safe vaccines and biomedical tools. The group of *Mononegavirales* represents an exceptional source of diverse viruses for diverse applications, for example, rabies virus neurotracing. Sendai virus, in particular, has emerged in the past years as a cornucopia for many novel and innovative biomedical tools, as impressively illustrated in the following chapters of this volume.

Acknowledgments Research in the author's laboratory is funded by the German Research Foundation (DFG) through SFB 870 and Grako 1202, and the German Federal Ministry of Education and Research (BMBF) grant no. 01KI1016B. I thank Kerstin Schuhmann and Yoshiyuki Nagai for valuable comments on the manuscript.

References

Ammayappan A, Lapatra SE, Vakharia VN (2010) A vaccinia-virus-free reverse genetics system for infectious hematopoietic necrosis virus. J Virol Methods 167:132–139

Astic L, Saucier D, Coulon P, Lafay F, Flamand A (1993) The CVS strain of rabies virus as trans-neuronal tracer in the olfactory system of mice. Brain Res 619:146–156

Baltimore D (1971) Expression of animal virus genomes. Bacteriol Rev 35:235–241

Baltimore D, Huang AS, Stampfer M (1970) Ribonucleic acid synthesis of vesicular stomatitis virus, II. An RNA polymerase in the virion. Proc Natl Acad Sci USA 66:572–576

Baron MD, Barrett T (1997) Rescue of rinderpest virus from cloned cDNA. J Virol 71:1265–1271

Biacchesi S (2011) The reverse genetics applied to fish RNA viruses. Vet Res 42:12

Blount KF, Uhlenbeck OC (2002) The hammerhead ribozyme. Biochem Soc Trans 30:1119–1122

Boonyaratanakornkit J, Bartlett E, Schomacker H, Surman S, Akira S, Bae YS, Collins P, Murphy B, Schmidt A (2011) The C proteins of human parainfluenza virus type 1 limit double-stranded RNA accumulation that would otherwise trigger activation of MDA5 and protein kinase R. J Virol 85:1495–1506

Brandler S, Ruffie C, Najburg V, Frenkiel MP, Bedouelle H, Despres P, Tangy F (2010) Pediatric measles vaccine expressing a dengue tetravalent antigen elicits neutralizing antibodies against all four dengue viruses. Vaccine 28:6730–6739

Bridgen A, Elliott RM (1996) Rescue of a segmented negative-strand RNA virus entirely from cloned complementary DNAs. Proc Natl Acad Sci USA 93:15400–15404

Bruns AM, Horvath CM (2012) Activation of RIG-I-like receptor signal transduction. Crit Rev Biochem Mol Biol 47:194–206

Brzózka K, Finke S, Conzelmann KK (2005) Identification of the rabies virus alpha/beta interferon antagonist: phosphoprotein P interferes with phosphorylation of interferon regulatory factor 3. J Virol 79:7673–7681

Brzózka K, Finke S, Conzelmann KK (2006) Inhibition of interferon signaling by rabies virus phosphoprotein P: activation-dependent binding of STAT1 and STAT2. J Virol 80:2675–2683

Buchholz UJ, Finke S, Conzelmann KK (1999) Generation of bovine respiratory syncytial virus (BRSV) from cDNA: BRSV NS2 is not essential for virus replication in tissue culture, and the human RSV leader region acts as a functional BRSV genome promoter. J Virol 73:251–259

Calain P, Roux L (1993) The rule of six, a basic feature for efficient replication of Sendai virus defective interfering RNA. J Virol 67:4822–4830

Cattaneo R, Miest T, Shashkova EV, Barry MA (2008) Reprogrammed viruses as cancer therapeutics: targeted, armed and shielded. Nat Rev Microbiol 6:529–540

Collins PL, Hill MG, Camargo E, Grosfeld H, Chanock RM, Murphy BR (1995) Production of infectious human respiratory syncytial virus from cloned cDNA confirms an essential role for the transcription elongation factor from the 5′ proximal open reading frame of the M2 mRNA in gene expression and provides a capability for vaccine development. Proc Natl Acad Sci USA 92:11563–11567

Conzelmann KK (1996) Genetic manipulation of non-segmented negative-strand RNA viruses. J Gen Virol 77(pt 3):381–389

Conzelmann KK (2004) Reverse genetics of Mononegavirales. Curr Top Microbiol Immunol 283:1–41

Conzelmann KK, Schnell M (1994) Rescue of synthetic genomic RNA analogs of rabies virus by plasmid-encoded proteins. J Virol 68:713–719

Cui S, Eisenacher K, Kirchhofer A, Brzózka K, Lammens A, Lammens K, Fujita T, Conzelmann KK, Krug A, Hopfner KP (2008) The C-terminal regulatory domain is the RNA 5′-triphosphate sensor of RIG-I. Mol Cell 29:169–179

Dum RP, Strick PL (2012) Transneuronal tracing with neurotropic viruses reveals network macro-architecture. Curr Opin Neurobiol doi:10.1016/j.conb.2012.12.002

Durbin AP, Hall SL, Siew JW, Whitehead SS, Collins PL, Murphy BR (1997) Recovery of infectious human parainfluenza virus type 3 from cDNA. Virology 235:323–332
Etessami R, Conzelmann KK, Fadai-Ghotbi B, Natelson B, Tsiang H, Ceccaldi PE (2000) Spread and pathogenic characteristics of a G-deficient rabies virus recombinant: an in vitro and in vivo study. J Gen Virol 81:2147–2153
Finke S, Conzelmann KK (1997) Ambisense gene expression from recombinant rabies virus: random packaging of positive- and negative-strand ribonucleoprotein complexes into rabies virions. J Virol 71:7281–7288
Finke S, Cox JH, Conzelmann KK (2000) Differential transcription attenuation of rabies virus genes by intergenic regions: generation of recombinant viruses overexpressing the polymerase gene. J Virol 74:7261–7269
Finke S, Mueller-Waldeck R, Conzelmann KK (2003) Rabies virus matrix protein regulates the balance of virus transcription and replication. J Gen Virol 84:1613–1621
Fitzgerald KA, McWhirter SM, Faia KL, Rowe DC, Latz E, Golenbock DT, Coyle AJ, Liao SM, Maniatis T (2003) IKKepsilon and TBK1 are essential components of the IRF3 signaling pathway. Nat Immunol 4:491–496
Ford E, Thanos D (2010) The transcriptional code of human IFN-beta gene expression. Biochim Biophys Acta 1799:328–336
Fu ZF (2005) Genetic comparison of the Rhabdoviruses from animals and plants. Curr Top Microbiol Immunol 292:1–24
Fuerst TR, Niles EG, Studier FW, Moss B (1986) Eukaryotic transient-expression system based on recombinant vaccinia virus that synthesizes bacteriophage T7 RNA polymerase. Proc Natl Acad Sci USA 83:8122–8126
Garcin D, Pelet T, Calain P, Roux L, Curran J, Kolakofsky D (1995) A highly recombinogenic system for the recovery of infectious Sendai paramyxovirus from cDNA: generation of a novel copy-back nondefective interfering virus. EMBO J 14:6087–6094
Geisbert TW, Feldmann H (2011) Recombinant vesicular stomatitis virus-based vaccines against Ebola and Marburg virus infections. J Infect Dis 204(Suppl 3):S1075–S1081
Gerlier D, Lyles DS (2011) Interplay between innate immunity and negative-strand RNA viruses: towards a rational model. Microbiol Mol Biol Rev 75:468–490
Ghanem A, Kern A, Conzelmann KK (2012) Significantly improved rescue of rabies virus from cDNA plasmids. Eur J Cell Biol 91:10–16
Ginger M, Haberl M, Conzelmann KK, Schwarz MK, Frick A (2013) Revealing the secrets of neuronal circuits with recombinant rabies virus technology. Front Neural Circuits 7:2. doi:10.3389/fncir.2013.00002, Epub@2013 Jan 24:2
Goff SP, Berg P (1976) Construction of hybrid viruses containing SV40 and lambda phage DNA segments and their propagation in cultured monkey cells. Cell 9:695–705
Gomme EA, Wanjalla CN, Wirblich C, Schnell MJ (2011) Rabies virus as a research tool and viral vaccine vector. Adv Virus Res 79:139–164
Goodbourn S, Randall RE (2009) The regulation of type I interferon production by paramyxoviruses. J Interferon Cytokine Res 29:539–547
He B, Paterson RG, Ward CD, Lamb RA (1997) Recovery of infectious SV5 from cloned DNA and expression of a foreign gene. Virology 237:249–260
Hoffman MA, Banerjee AK (1997) An infectious clone of human parainfluenza virus type 3. J Virol 71:4272–4277
Honda K, Takaoka A, Taniguchi T (2006) Type I interferon [corrected] gene induction by the interferon regulatory factor family of transcription factors. Immunity 25:349–360
Hornung V, Ellegast J, Kim S, Brzózka K, Jung A, Kato H, Poeck H, Akira S, Conzelmann KK, Schlee M, Endres S, Hartmann G (2006) 5′-Triphosphate RNA is the ligand for RIG-I. Science 314:994–997
Huang Y, Tang Q, Nadin-Davis SA, Zhang S, Hooper CD, Ming P, Du J, Tao X, Hu R, Liang G (2010) Development of a reverse genetics system for a human rabies virus vaccine strain employed in China. Virus Res 149:28–35
Inoue K, Shoji Y, Kurane I, Iijima T, Sakai T, Morimoto K (2003) An improved method for recovering rabies virus from cloned cDNA. J Virol Methods 107:229–236

Ito N, Moseley GW, Blondel D, Shimizu K, Rowe CL, Ito Y, Masatani T, Nakagawa K, Jans DA, Sugiyama M (2010) The role of interferon-antagonist activity of rabies virus phosphoprotein in viral pathogenicity. J Virol 84:6699–6710

Johnson JE, Schnell MJ, Buonocore L, Rose JK (1997) Specific targeting to CD4+ cells of recombinant vesicular stomatitis viruses encoding human immunodeficiency virus envelope proteins. J Virol 71:5060–5068

Kato A, Sakai Y, Shioda T, Kondo T, Nakanishi M, Nagai Y (1996) Initiation of Sendai virus multiplication from transfected cDNA or RNA with negative or positive sense. Genes Cells 1:569–579

Kato H, Takahasi K, Fujita T (2011) RIG-I-like receptors: cytoplasmic sensors for non-self RNA. Immunol Rev 243:91–98

Kawasaki T, Kawai T, Akira S (2011) Recognition of nucleic acids by pattern-recognition receptors and its relevance in autoimmunity. Immunol Rev 243:61–73

Khattar SK, Samal S, Devico AL, Collins PL, Samal SK (2011) Newcastle disease virus expressing human immunodeficiency virus type 1 envelope glycoprotein induces strong mucosal and serum antibody responses in Guinea pigs. J Virol 85:10529–10541

Kimura Y (1973) Phenotypic mixing of vesicular stomatitis virus with HVJ (Sendai virus). Jpn J Microbiol 17:373–381

Kinoh H, Inoue M, Komaru A, Ueda Y, Hasegawa M, Yonemitsu Y (2009) Generation of optimized and urokinase-targeted oncolytic Sendai virus vectors applicable for various human malignancies. Gene Ther 16:392–403

Klingen Y, Conzelmann KK, Finke S (2008) Double-labeled rabies virus: live tracking of enveloped virus transport. J Virol 82:237–245

Lafon M (2011) Evasive strategies in rabies virus infection. Adv Virus Res 79:33–53

Lamb RA (2007) Mononegavirales. In: Knipe DM, Howley PM (eds) Fields virology. Lippincott, Philadelphia, pp 1357–1362

Lawson ND, Stillman EA, Whitt MA, Rose JK (1995) Recombinant vesicular stomatitis viruses from DNA. Proc Natl Acad Sci USA 92:4477–4481

Le Mercier P, Garcin D, Hausmann S, Kolakofsky D (2002) Ambisense Sendai viruses are inherently unstable but are useful to study viral RNA synthesis. J Virol 76:5492–5502

Loo YM, Gale M Jr (2011) Immune signaling by RIG-I-like receptors. Immunity 34:680–692

Luytjes W, Krystal M, Enami M, Pavin JD, Palese P (1989) Amplification, expression, and packaging of foreign gene by influenza virus. Cell 59:1107–1113

Marriott AC, Easton AJ (1999) Reverse genetics of the Paramyxoviridae. Adv Virus Res 53:321–340

Marschalek A, Finke S, Schwemmle M, Mayer D, Heimrich B, Stitz L, Conzelmann KK (2009) Attenuation of rabies virus replication and virulence by picornavirus internal ribosome entry site elements. J Virol 83:1911–1919

Martin A, Staeheli P, Schneider U (2006) RNA polymerase II-controlled expression of antigenomic RNA enhances the rescue efficacies of two different members of the Mononegavirales independently of the site of viral genome replication. J Virol 80:5708–5715

Mebatsion T (2001) Extensive attenuation of rabies virus by simultaneously modifying the dynein light chain binding site in the P protein and replacing Arg333 in the G protein. J Virol 75:11496–11502

Mebatsion T, Conzelmann KK (1996) Specific infection of CD4+ target cells by recombinant rabies virus pseudotypes carrying the HIV-1 envelope spike protein. Proc Natl Acad Sci USA 93:11366–11370

Mebatsion T, Konig M, Conzelmann KK (1996) Budding of rabies virus particles in the absence of the spike glycoprotein. Cell 84:941–951

Mebatsion T, Finke S, Weiland F, Conzelmann KK (1997) A CXCR4/CD4 pseudotype rhabdovirus that selectively infects HIV-1 envelope protein-expressing cells. Cell 90:841–847

Ming PG, Huang Y, Tang Q, Du JL, Tao XY, Yan JX, Hu RL (2009) Construction and analysis of full-length cDNA clone of rabies virus street strain. Bing Du Xue Bao 25:17–22

Morodomi Y, Yano T, Kinoh H, Harada Y, Saito S, Kyuragi R, Yoshida K, Onimaru M, Shoji F, Yoshida T, Ito K, Shikada Y, Maruyama R, Hasegawa M, Maehara Y, Yonemitsu Y (2012) BioKnife, a uPA activity-dependent oncolytic Sendai virus, eliminates pleural spread of malignant mesothelioma via simultaneous stimulation of uPA expression. Mol Ther 20:769–777

Motz C, Schuhmann KM, Kirchhofer A, Moldt M, Witte G, Conzelmann KK, Hopfner KP (2013) Paramyxovirus V proteins disrupt the fold of the RNA sensor MDA5 to inhibit antiviral signaling. Science 339:690–693
Mourez T, Mesel-Lemoine M, Combredet C, Najburg V, Cayet N, Tangy F (2011) A chimeric measles virus with a lentiviral envelope replicates exclusively in CD4+/CCR5+ cells. Virology 419:117–125
Muik A, Kneiske I, Werbizki M, Wilfingseder D, Giroglou T, Ebert O, Kraft A, Dietrich U, Zimmer G, Momma S, von Laer D (2011) Pseudotyping vesicular stomatitis virus witn lymphocytic choriomeningitis virus glycoproteins enhances infectivity for glioma cells and minimizes neurotropism. J Virol 85:5679–5684
Nagai Y, Takakura A, Irie T, Yonemitsu Y, Gotoh B (2011) Sendai virus: evolution from mouse pathogen to a state-of-the-art tool in virus research and biotechnology. In: Samal SK (ed) The biology of paramyxoviruses. Caister Academic, Norfolk, pp 115–173
Nishimura K, Sano M, Ohtaka M, Furuta B, Umemura Y, Nakajima Y, Ikehara Y, Kobayashi T, Segawa H, Takayasu S, Sato H, Motomura K, Uchida E, Kanayasu-Toyoda T, Asashima M, Nakauchi H, Yamaguchi T, Nakanishi M (2011) Development of defective and persistent Sendai virus vector: a unique gene delivery/expression system ideal for cell reprogramming. J Biol Chem 286:4760–4771
Onoguchi K, Yoneyama M, Takemura A, Akira S, Taniguchi T, Namiki H, Fujita T (2007) Viral infections activate types I and III interferon genes through a common mechanism. J Biol Chem 282:7576–7581
Orbanz J, Finke S (2010) Generation of recombinant European bat lyssavirus type 1 and inter-genotypic compatibility of lyssavirus genotype 1 and 5 antigenome promoters. Arch Virol 155:1631–1641
Park KH, Huang T, Correia FF, Krystal M (1991) Rescue of a foreign gene by Sendai virus. Proc Natl Acad Sci USA 88:5537–5541
Pattnaik AK, Wertz GW (1990) Replication and amplification of defective interfering particle RNAs of vesicular stomatitis virus in cells expressing viral proteins from vectors containing cloned cDNAs 42. J Virol 64:2948–2957
Pattnaik AK, Wertz GW (1991) Cells that express all five proteins of vesicular stomatitis virus from cloned cDNAs support replication, assembly, and budding of defective interfering particles. Proc Natl Acad Sci USA 88:1379–1383
Pattnaik AK, Ball LA, LeGrone AW, Wertz GW (1992) Infectious defective interfering particles of VSV from transcripts of a cDNA clone. Cell 69:1011–1020
Perrotta AT, Been MD (1990) The self-cleaving domain from the genomic RNA of hepatitis delta virus: sequence requirements and the effects of denaturant. Nucleic Acids Res 18:6821–6827
Pfaller CK, Radeke MJ, Cattaneo R, Samuel CE (2013) Measles virus C protein impairs production of defective copyback double-stranded viral RNA and activation of protein kinase R. J Virol. Oct 23. [Epub ahead of print] PubMed PMID: 24155404
Platanias LC (2005) Mechanisms of type-I- and type-II-interferon-mediated signalling. Nat Rev Immunol 5:375–386
Pringle CR (2005) Mononegavirales. In: Fauquet CM, Mayo MA, Maniloff J, Desselberger U, Ball LA (eds) Virus taxonomy. Eighth report of the International Committee on the taxonomy of viruses. Elsevier/Academic, London, pp 609–614
Racaniello VR, Baltimore D (1981) Cloned poliovirus complementary DNA is infectious in mammalian cells. Science 214:916–919
Radecke F, Spielhofer P, Schneider H, Kaelin K, Huber M, Dotsch C, Christiansen G, Billeter MA (1995) Rescue of measles viruses from cloned DNA. EMBO J 14:5773–5784
Rajani KR, Pettit Kneller EL, McKenzie MO, Horita DA, Chou JW, Lyles DS (2012) Complexes of vesicular stomatitis virus matrix protein with host Rae1 and Nup98 involved in inhibition of host transcription. PLoS Pathog 8:e1002929
Randall RE, Goodbourn S (2008) Interferons and viruses: an interplay between induction, signalling, antiviral responses and virus countermeasures. J Gen Virol 89:1–47

Rieder M, Conzelmann KK (2011) Interferon in rabies virus infection. Adv Virus Res 79:91–114

Rieder M, Brzózka K, Pfaller CK, Cox JH, Stitz L, Conzelmann KK (2011) Genetic dissection of interferon-antagonistic functions of rabies virus phosphoprotein: inhibition of interferon regulatory factor 3 activation is important for pathogenicity. J Virol 85:842–852

Roberts A, Rose JK (1998) Recovery of negative-strand RNA viruses from plasmid DNAs: a positive approach revitalizes a negative field. Virology 247:1–6

Rothenberg E, Smotkin D, Baltimore D, Weinberg RA (1977) In vitro synthesis of infectious DNA of murine leukaemia virus. Nature (Lond) 269:122–126

Russell SJ, Peng KW, Bell JC (2012) Oncolytic virotherapy. Nat Biotechnol 30:658–670

Sakaguchi T, Kato A, Kiyotani K, Yoshida T, Nagai Y (2008) Studies on the paramyxovirus accessory genes by reverse genetics in the Sendai virus-mouse system. Proc Jpn Acad Ser B Phys Biol Sci 84:439–451

Schnell MJ, Mebatsion T, Conzelmann KK (1994) Infectious rabies viruses from cloned cDNA. EMBO J 13:4195–4203

Schnell MJ, Buonocore L, Kretzschmar E, Johnson E, Rose JK (1996) Foreign glycoproteins expressed from recombinant vesicular stomatitis viruses are incorporated efficiently into virus particles. Proc Natl Acad Sci USA 93:11359–11365

Schnell MJ, Johnson JE, Buonocore L, Rose JK (1997) Construction of a novel virus that targets HIV-1-infected cells and controls HIV-1 infection. Cell 90:849–857

Schnell MJ, Buonocore L, Boritz E, Ghosh HP, Chernish R, Rose JK (1998) Requirement for a non-specific glycoprotein cytoplasmic domain sequence to drive efficient budding of vesicular stomatitis virus. EMBO J 17:1289–1296

Schoggins JW, Wilson SJ, Panis M, Murphy MY, Jones CT, Bieniasz P, Rice CM (2011) A diverse range of gene products are effectors of the type I interferon antiviral response. Nature (Lond) 472:481–485

Schuhmann KM, Pfaller CK, Conzelmann KK (2011) The measles virus V protein binds to p65 (RelA) to suppress NF-{kappa}B activity. J Virol 85:3162–3171

Sharma S, tenOever BR, Grandvaux N, Zhou GP, Lin R, Hiscott J (2003) Triggering the interferon antiviral response through an IKK-related pathway. Science 300:1148–1151

Sharmeen L, Kuo MY, Dinter-Gottlieb G, Taylor J (1988) Antigenomic RNA of human hepatitis delta virus can undergo self-cleavage. J Virol 62:2674–2679

Sparrer KM, Pfaller CK, Conzelmann KK (2012) Measles virus C protein interferes with beta interferon transcription in the nucleus. J Virol 86:796–805

Taniguchi T, Palmieri M, Weissmann C (1978) QB DNA-containing hybrid plasmids giving rise to QB phage formation in the bacterial host. Nature (Lond) 274:223–228

Tao L, Ge J, Wang X, Zhai H, Hua T, Zhao B, Kong D, Yang C, Chen H, Bu Z (2010) Molecular basis of neurovirulence of flury rabies virus vaccine strains: importance of the polymerase and the glycoprotein R333Q mutation. J Virol 84:8926–8936

Theofilopoulos AN, Baccala R, Beutler B, Kono DH (2005) Type I interferons (alpha/beta) in immunity and autoimmunity. Annu Rev Immunol 23:307–335

Theriault S, Groseth A, Artsob H, Feldmann H (2005) The role of reverse genetics systems in determining filovirus pathogenicity. Arch Virol Suppl 19:157–177

Ugolini G (1995) Specificity of rabies virus as a transneuronal tracer of motor networks: transfer from hypoglossal motoneurons to connected second-order and higher order central nervous system cell groups. J Comp Neurol 356:457–480

Versteeg GA, Garcia-Sastre A (2010) Viral tricks to grid-lock the type I interferon system. Curr Opin Microbiol 13:508–516

Vulliemoz D, Roux L (2001) "Rule of six:" how does the Sendai virus RNA polymerase keep count? J Virol 75:4506–4518

Weissmann C, Weber H, Taniguchi T, Muller W, Meyer F (1979) Reversed genetics: a new approach to the elucidation of structure–function relationship. Ciba Found Symp 66:47–61

Whelan SP, Ball LA, Barr JN, Wertz GT (1995) Efficient recovery of infectious vesicular stomatitis virus entirely from cDNA clones. Proc Natl Acad Sci USA 92:8388–8392

Whelan SP, Barr JN, Wertz GW (2004) Transcription and replication of nonsegmented negative-strand RNA viruses. Curr Top Microbiol Immunol 283:61–119

Wickersham IR, Feinberg EH (2012) New technologies for imaging synaptic partners. Curr Opin Neurobiol 22:121–127

Wickersham IR, Lyon DC, Barnard RJ, Mori T, Finke S, Conzelmann KK, Young JA, Callaway EM (2007) Monosynaptic restriction of transsynaptic tracing from single, genetically targeted neurons. Neuron 53:639–647

Yoneyama M, Suhara W, Fukuhara Y, Fukuda M, Nishida E, Fujita T (1998) Direct triggering of the type I interferon system by virus infection: activation of a transcription factor complex containing IRF-3 and CBP/p300. EMBO J 17:1087–1095

Yu S, Feng X, Shu T, Matano T, Hasegawa M, Wang X, Li H, Li Z, Zhong R, Zeng Y (2010) Comparison of the expression and immunogenicity of wild-type and sequence-modified HIV-1 gag genes in a recombinant Sendai virus vector. Curr HIV Res 8:199–206

Chapter 2
Sendai Virus Biology and Engineering Leading up to the Development of a Novel Class of Expression Vector

Yoshiyuki Nagai and Atsushi Kato

Abstract Sendai virus (SeV), a prototypic member of the family *Paramyxoviridae*, was discovered in 1953, six decades ago. It is not just an old mouse pathogen but has been an irreplaceable model in in basic research to understand paramyxovirus replication and pathogenesis. The SeV reverse genetics established in 1996 has played a particularly prominent role in this context by settling outstanding issues and resolving enigmas. At the same time, the technology is evolving into a multipurpose cytoplasmic (nonintegrating) RNA vector. Its diverse medical applications are now in the pipeline and being tested in clinical settings as illustrated in the subsequent chapters. The production of diverse target-oriented devices has been possible by making full use of a variety of SeV theories and traits discovered during the six decades. Here, we summarize the long journey of SeV research leading up to the invention of this novel class of expression vector, SeV vector.

2.1 Introduction

Sendai virus (SeV), also named murine parainfluenza virus 1, was discovered nearly six decades ago. It is an enveloped virus with a nonsegmented single molecule of linear, negative sense RNA genome, hence belonging to the order (superfamily) *Mononegavirales*. More specifically, SeV is the type species of the genus *Respirovirus* in the subfamily *Pramyxovirinae* of the family *Paramyxoviridae*.

Y. Nagai (✉)
RIKEN Center of Research Network for Infectious Diseases, Jimbocho 101 Bldg. 8F, Kandajimbo-cho 1-101, Chiyoda-ku, Tokyo 101-0051, Japan
e-mail: yoshi-nagai@riken.jp

A. Kato
Division of Radiological Protection & Biology and Division of Quality Assurance, National Institute of Infectious Diseases, Toyama 1-23-1, Shinjyuku-ku, Tokyo 162-8640, Japan

Y. Nagai (ed.), *Sendai Virus Vector: Advantages and Applications*,
DOI 10.1007/978-4-431-54556-9_2, © Springer Japan 2013

This family comprises a wide range of "classic" pathogens of medical or veterinary significance such as parainfluenza, measles, mumps, respiratory syncytial, canine distemper and Newcastle disease viruses and "newly emergent" deadly agents such as Nipah and Hendra viruses. On the other hand, no causal relationship between SeV and any human or animal diseases thus far has been demonstrated except that it has posed a serious threat to the rodents in animal research facilities. Thus, SeV may be a virus that eventually disappears from the virological scene. Nevertheless, it remains a major player in virology because SeV has been one of the best-studied viruses in the *Paramyxoviridae* to understand the mechanism of replication and pathogenesis of this virus family and is now attracting a wide attention as a novel class of expression vector with enormous versatility. This chapter summarizes the history of SeV research, focusing on the establishment of its reverse genetics that provided the technological basis of SeV vector development for medical use in general, and various biological aspects of SeV that have facilitated the production of target-oriented devices to be used in special settings.

2.2 Discovery and Origin of SeV

SeV was first isolated in 1953 at Tohoku University Hospital in Sendai, Japan, as an agent causative of an outbreak of fatal pneumonia of newborn children (Kuroya et al. 1953). The materials used for virus isolation were the homogenates of autopsy lung tissues taken from the victims of the disease, which were intranasally inoculated into mice. The inoculated mice developed influenza-like lesions in the lungs. The homogenates of those mouse lung tissues were then transferred to the amniotic cavities of chicken embryonated eggs, and the amniotic fluid harvested several days later displayed a high level of agglutination of chicken erythrocytes. Because the laboratory tests available for identification of known influenza and other hemagglutinating viruses were all negative, the isolate was concluded to be a new virus (reviewed in Ishida and Homma 1978).

However, subsequent studies have so far raised serious doubts concerning the causality between SeV and the children's pneumonia. For example, Fukumi et al. (1954) became aware that passaging in mouse lungs of influenza virus isolates from various regions of Japan quite often generated a virus antigenically totally different from the original influenza virus. This new agent was later identified as SeV. It was even said that dropping physiological saline alone into the noses of apparently healthy mice could sometimes activate the silently infecting SeV, resulting in induction of severe pneumonia. These observations suggested nationwide enzootic (subclinical) spread of SeV among mice during the 1950s. The virus appeared to be readily activated by various exogenous stimuli to the airway of the mouse. In the situation where the virus could not be regarded as a human pathogen, the Japanese Society for Virology decided to name it hemagglutinating virus of Japan (HVJ) (Ishida and Homma 1978; Nagai et al. 2011).

Indeed, spontaneous outbreaks of SeV infection have been a devastating threat to mice and rats in animal experimental facilities worldwide. Extensive national serological surveys of SeV and other microbes have been conducted for mouse and rat colonies of many animal facilities belonging to universities, pharmaceutical industries, and animal breeders in Japan since 1980. Contamination of mouse and rat colonies with SeV was found in as many as 25.0 % and 31.8 %, respectively, of the total colonies tested in 1980 (Nagai et al. 2011). The positive rate decreased year by year and became zero for both mouse and rat colonies in 2009, through introduction of barrier systems into facilities, which had previously maintained animals under open and conventional conditions, and careful monitoring of animals to be brought in and maintained. Similar serological monitoring studies also revealed decreasing trends of SeV contamination in several other countries (Nagai et al. 2011).

Easterbrook et al. (2008) reported that 4.1 % of a total of 162 wild-caught Norway rats were SeV antibody positive, suggesting that SeV has originated from rodents. This view was substantiated by a more recent study demonstrating the presence of viral sequences conspecific with SeV in wild rodents (*Rattus rattus*) captured in Thailand by reverse transcription-polymerase chain reaction (RT-PCR) targeting the viral *L* gene (Drexler et al. 2012).

2.3 Structure of the Virion

Figure 2.1 shows a negatively stained SeV particle (virion) (a) and its schematic diagram (b). The SeV particle (virion) consists of three major structural components: the inner ribonucleoprotein complex (RNP), the outer envelope, and the matrix (M) protein layer between the internal and external structures (Fig. 2.1b). The ultrastructure has been revealed mainly using egg-grown virions. It is spherical, with a diameter of approximately 200 nm. With freshly prepared, intact virions, which were harvested from eggs relatively early (~24 h) after infection and immediately processed for negative staining, uranyl acetate did not penetrate through the envelope, and no RNP was visible. The internal RNP structure can be seen easily using virions with the envelopes partially damaged by late harvesting (at ~72 h) (Fig. 2.1a), freezing and thawing, or storage in a refrigerator (at 4 °C) for a considerable period (1 week) (Shimizu et al. 1976; reviewed in Nagai et al. 2011).

2.3.1 RNP

The RNP is best studied after isolation from detergent-solubilized virions. The RNP consists of the RNA genome that is encapsidated with the nucleocapsid (N) proteins. It has a left-handed helical structure 1.0 μm in length and 18 nm in diameter, with a helical pitch of about 5 nm and a central hollow core of 4 nm, and contains

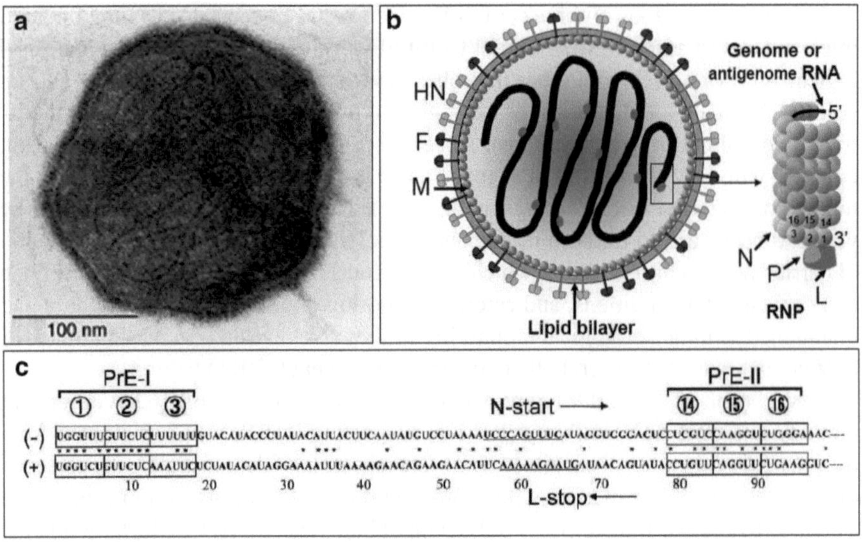

Fig. 2.1 A negatively stained SeV particle (courtesy of Y. Ueda) (**a**) and schematic diagram of an SeV particle (**b**). The N-protein units covering the hexamers consisting of two separate promoters are positioned on the same surface of the helical RNP (**b**). For details, see text. (Adapted from Nagai et al. 2011, with permission.) (**c**) The 3′-terminal sequences of SeV genome (–) and antigenome (+). *Asterisks* indicate nucleotide identity between the genome and antigenome. The hexamers constituting the two separate promoters, PrE-I and PrE-II, are *boxed* (Tapparel et al. 1998). (Adapted from Nagai et al. 2011, with permission)

13 N proteins per turn (Hosaka 1968; Finch and Gibbs 1970; Egelman et al. 1989; Fig. 2.1b). Because one N protein is thought to bind to every six RNA bases (Egelman et al. 1989; Calain and Roux 1993) and because the SeV genome consists of 15,384 nucleotides (nt), a single RNP is estimated to contain 2,564 units of N protein. However, SeV RNP is flexible in structure and exists in a few different helical states with a pitch of 5.3, 6.8, or 37.5 nm (Egelman et al. 1989; Heggeness et al. 1982; reviewed in Lamb and Parks 2007).

SeV particles are pleomorphic. This nature was first described by Hosaka et al. (1966). By negative staining, the authors analyzed the sizes and RNP copy number in SeV particles fractionated by sucrose density-gradient centrifugation into three bands that contained, respectively, the particles of different sizes. Overall, the size ranged from 160 to 600 nm. Depending on the size, the virions contained one to several units of RNP. The multiple RNP strands looked continuous rather than separated, reaching two to several micrometers or more in length. If the viruses contained one unit of 1-μm length RNP, they were infectious, indicating that 1 μm RNP could represent a minimum infectious unit. The infectivity of this type of virions was inactivated by ^{60}Co irradiation with linear, single-hit kinetics, whereas those containing multiple RNP copies displayed nonlinear, multiple-hit kinetics with a shoulder. Thus, the multiple RNPs could represent multiple infectious units.

Essentially the same results were obtained by a recent cryo-electron tomography, showing that the SeV particle sizes ranged from 110 to 540 nm in diameter and that one to six RNPs were packaged in a virion (Loney et al. 2009). The foregoing study by Hosaka et al. (1966) further pointed out that the particles containing multiple RNPs accounted for only 20 % of the total. Similarly, another study examined quantitatively the size distribution of virus particles in an SeV stock and showed a normal distribution with a sharp peak at 175–224 nm in diameter, and 68.27 % of the virions were within the range of 200±48 nm (Sakaguchi et al. 2002) (see Sect. 2.5.7.2). In addition, for another paramyxovirus Newcastle disease virus (NDV) Goff et al. (2012) generated three different recombinant virus clones, each expressing a unique fluorescent protein and showed that a majority of virions contained one RNP, whereas approximately 25 % had two or more functional RNPs. Overall, although pleomorphic, SeV and paramyxovirus morphogenesis still appears to be operated under a considerable control mechanism.

The paramyxovirus L (large) and P (phospho) proteins together constitute the RNA polymerase complex (Hamaguchi et al. 1983; Horikami et al. 1992) and are bound to the RNP. Assuming that about 300 P and 50 L proteins are attached to each SeV RNP (Lamb et al. 1976) and that a P protein tetramer and one L protein constitute the active L–P complex, there appear to be, at most, 50 units of active polymerase on each RNP template. Using immunogold labeling the P proteins were found to be dispersed uniformly along the SeV RNPs isolated from the virions, whereas they were present in discrete clusters at varying locations of the RNPs isolated from the cytosol (Portner and Murti 1986). The L proteins were colocalized with those P protein clusters, suggesting that the P and L are comobile along the intracellular RNP to act as the RNA polymerase (Portner et al. 1988). The isolated RNPs are sometimes broader (40 nm) in diameter, which appeared to be the result of additional binding of the M proteins. The broader RNP was abundant in the cytosol, suggesting the role of M protein in the recognition between the RNP and the envelope proteins in the process of virus assembly (Portner and Murti 1986; reviewed in Nagai et al. 2011).

2.3.2 *Envelope Glycoproteins and M Protein*

The outermost structure, the envelope, contains two types of glycoproteins protruding in the form of spikes, anchored within the lipid bilayer derived from the plasma membrane of host cells (Fig. 2.1b). One of them has hemagglutination (HA) and neuraminidase (NA) activities (Tozawa et al. 1973) and is named HN. It is a type II integral membrane protein, spanning the membrane once and possessing an N-terminal cytoplasmic tail, a single transmembrane domain, a membrane proximal stalk domain, and a large C-terminal globular domain, and is present on the virion surface as a tetramer consisting of a pair of disulfide-bonded dimers. Although the stalk domain is important for tetramer formation, the head domain contains the sites for HA activity and NA activity (Lamb and Parks 2007). It has been an issue of debate for years whether paramyxovirus HN protein possesses combined or separate sites for HA and NA activities. The fact that a compound named BCX-2798 inhibited NA activity but not HA activity of SeV suggested that the two active sites are separated (Bousse and Takimoto 2006). The SeV HN protein possesses four potential *N*-glycosylation sites. Oligosaccharide chains are actually attached to three of them but not to the remaining one (Segawa et al. 2003) (see Sect. 2.5.6.2). The HN protein recognizes terminal sialic acids in the sugar chains as the receptor on the target cell surface. It also contributes to the release of progeny virus from the cell surface (Nagai et al. 2011) (see Sects. 2.5.1.1 and 2.5.7.5).

The other spike glycoprotein is called F protein because it induces fusion of the viral envelope with the host cell plasma membrane, thereby introducing the viral RNP into the cytoplasm. It is a type I transmembrane glycoprotein, which possesses the N-terminal signal peptide and the C-terminal membrane anchoring domain, leaving a short cytoplasmic tail. All three potential *N*-glycosylation sites on the SeV F protein were shown to be glycosylated (Segawa et al. 2000) (see Sect. 2.5.6.2). It is synthesized as an inactive precursor F_0. The biologically active F is generated through posttranslational cleavage between the Arg116 and Phe117 of the F_0 into two disulfide-linked subunits, the N-terminal (extracellular) F_2 and C-terminal (transmembrane) F_1 subunits (Homma and Ouchi 1973; Scheid and Choppin 1974) (see Sect. 2.5.6.3). Cleavage releases a hydrophobic fusion peptide (26 residues) at the N-terminus of F_1 (Gething et al. 1978). The Cys70, the only cysteine residue in the F_2, and Cys199, the most upstream cysteine in the F_1, are involved in the inter-chain bond (Iwata et al. 1994) (see Sect. 2.5.6.2). The structural motif, consisting of a repeating pattern of seven hydrophobic amino acids, (4-3) heptad repeats A (HRA) and HRB, was identified just after the fusion peptide and just before the transmembrane domain, respectively. On the virion surface the F proteins exist as a homotrimer (Lamb and Parks 2007).

The M protein located between the envelope and RNP is thought to serve as an anchor to stabilize the spike glycoprotein molecules floating in the lipid bilayer by binding to their cytoplasmic tails; it also binds to RNP, crosslinking the internal and external structures (Shimizu and Ishida 1975; Yoshida et al. 1976; reviewed in Nagai et al. 2011) (see Sect. 2.5.7.2).

2.4 Nucleotide Sequencing and the Establishment of Reverse Genetics

2.4.1 The Nucleotide Sequence of the SeV Genome

The nucleotide sequencing of the entire paramyxovirus genome was first achieved for SeV in the mid-1980s (Shioda et al. 1983, 1986; Hidaka et al. 1984). The genome consists of 15,384 nucleotides (nt) with the 3′-extracistronic region, the leader sequence (55 nt) (*le*), and the 5′-extracistronic region, the trailer sequence (57 nt) (*tr*). The genes coding for the afore-described six structural proteins lie in the order of 3′-*N*-*P*-*M*-*F*-*HN*-*L*-5′. Each gene begins with the 10-nt-long transcription start signal (3′-UCCCNNUUNC-5′) and ends with the 11-nt-long transcription stop signal (3′-UNAUUCUUUUU-5′). The stop signal of the upstream gene is connected to the start signal of the gene just downstream with an intergenic 3 nucleotides (3′-GNN-5′). The lengths of the encoded proteins in the numbers of amino acids (aa) are 524 (N), 568 (P), 348 (M), 565 (F), 575 (HN), and 2,228 (L). The *P* gene further encodes two accessory proteins, the V (384 aa) protein and a nested set of four C proteins (215–175 aa) (see Sect. 2.5.5.1). Thus, the entire genome is now formulized as 3′*le-N-P/C/V-M-F-HN-L-tr*5′ (reviewed in Nagai et al. 2011).

2.4.2 The Establishment of Reverse Genetics as an Earnest Desire in the Post-Genome Era of Paramyxovirus Research

Plasmid-based expression of individual genes has enhanced our understanding of paramyxovirus protein functions. However, mere summation of the functions of individual genes does not guarantee an accurate overall picture of the virus, a viable organism in which various gene functions are intertwined with one another in a complex and systematic manner. Forward genetics fell short of providing methods of reproducing (and thus confirming) the phenotypes identified as corresponding to certain genotypes. Furthermore, there were a number of unsolved issues in paramyxovirus biology that had been unapproachable by conventional virology, including gene expression studies. Typical of such problems are the accessory genes encoding the V and C proteins because the virological definition of an accessory gene is that the gene can be deleted (knocked out) without totally destroying infectivity at the cellular level and because the actual roles of the accessory gene can be defined by comparing the phenotypes of the knockout virus with those of the parental virus at the cellular level and in the susceptible host organism (Nagai 1999; Nagai and Kato 2004; Nagai et al. 2011).

Under these circumstances, post-genome research of SeV and other nonsegmented negative strand RNA viruses (NNSVs) belonging to the order *Mononegavirales* looked forward to the establishment of reverse genetics, which would enable the recovery of infectious virus entirely from cDNA and allow gene manipulation at will and assessment of its outcome not only in terms of cellular-level proliferation but also in the context of viral pathogenesis at the level of the entire organism. In case of positive strand RNA viruses, such as poliovirus, recovery of virus from cDNA is relatively simple because the RNA genome intracellularly expressed from transfected cDNA or *in vitro* synthesized from cDNA and transfected into cells immediately acts as the mRNA and is capable of initiating the lifecycle. The naked genome of many DNA viruses can also initiate the lifecycle upon transfection. In contrast, for NNSVs, nothing happens when the RNA genome of either negative sense or antigenomic positive sense is expressed from the respective cDNAs in cells or when *in vitro* synthesized negative or positive sense genomic RNA is transfected. To initiate the lifecycle, a set of supporting proteins have to be coexpressed under fine-tuned conditions to generate a functional ribonucleoprotein complex RNP capable of initiating the lifecycle (see Sect. 2.5.3). Therefore, NNSV recovery entirely from cDNA required quite a complex experimental protocol and extensive efforts with trials and errors that continued over a decade. In 1994, at the end of a decade of trial-and-error experimentation, Conzelmann and his coworkers succeeded for the first time in recovery of an NNSV, rabies virus (in the family *Rhabdoviridae*), entirely from cDNA (Schnell et al. 1994) (see Chap. 1). Shortly thereafter, six groups across the globe, including ours, achieved rescue from cDNA of SeV, measles virus, and respiratory syncytial virus (RSV) in the *Paramyxoviridae* and another rhabodovirus vesicular stomatitis virus (VSV) (reviewed in Nagai 1999). By the end of the 1990s, the rescue system had been established for most paramyxoviruses, and hence paramyxovirus reverse genetics came of age (Nagai and Kato 1999).

2.4.3 The Protocol for SeV Recovery Entirely from cDNA

We assembled the 11 cDNA fragments spanning the entire genome of SeV strain Z, which had been used for nucleotide sequencing as described here, into a single chain and inserted it between the T7 promoter and the region encoding the hepatitis delta virus (HDV) ribozyme (Rbz) in a plasmid. The protocol of SeV recovery from this cDNA plasmid essentially followed that established for rabies virus (Schnell et al. 1994; see Chap. 1) with several modifications (Kato et al. 1996). Briefly, four plasmids encoding the entire genome, the N protein, P protein, and L protein were transfected into cells and driven by bacteriophage T7 polymerase expressed from vaccinia virus vector vTF7-3 (Fuerst et al. 1986). HDV Rbz placed immediately downstream of the viral sequence acted to generate the correct 3′-end of the full-length viral RNA. In addition, two inhibitors, rifampicin and arabinocylcytosine, were added to the culture medium to minimize the cytopathic effect of vTF7-3.

The efficiency of virus recovery was one infectious unit from 10^4 to 10^5 transfected cells in our system (Kato et al. 1996), whereas the efficiency was 50- to 100 fold lower in another SeV system (Garcin et al. 1995), which used the same full-length cDNA plasmid constructed by us, and in others established those days for a variety of NNSVs including the rabies virus setup that opened the era of NNSV reverse genetics (Schnell et al. 1994) (Nagai 1999). Perhaps because of this high efficiency, our SeV system could initiate SeV multiplication from transfection with all possible versions, that is, not only the cDNA plasmid directing either negative or positive sense RNA but also the RNA of positive or negative sense that had been generated *in vitro* from the respective cDNA plasmids (Kato et al. 1996). In all other setups established in those days, virus recovery appeared to be successful only from the cDNAs directing antigenomic RNA (Nagai 1999).

The remarkably high virus rescue rates in our SeV system probably resulted from fine tuning of the system (Palese et al. 1996). Particularly important could be constructing plasmids giving rise to full-length viral RNA with the 5′-end identical to that of the wild-type SeV RNA without extra G residues templated by the nonviral T7 promoter sequence so that the expressed RNA meets the rule of six (see Sect. 2.5.3.2) (Kato et al. 1996; reviewed in Nagai 1999). Indeed, three such nonviral Gs attached to the 5′-end of antigenomic RNA appeared to break the rule of six and be deleterious for SeV recovery, as no appreciable titer was recovered until they were deleted to create the correct 3′-end of genomic RNA during serial passages of the recovered virus in eggs (Garcin et al. 1995). Later, however, Leyrer et al. (1998) also established a rapid and efficient SeV recovery system from cDNA and indicated that if T7 promoter activity is strong enough, the resulting large amounts of transcripts can outweigh the inhibitory effect of the extra three Gs for replication.

Psoralen compound covalently binds to double-stranded DNA under UV light. UV irradiation of vTF7-3 in the presence of psoralen under certain conditions caused reduction of the vaccinia virus-associated cytopathogenicty, whereas T7 RNA polymerase expression remained little changed. As a result, recovery efficiency increased by 100- to 1,000 fold (1 infectious unit from 10^2 to 10^3 transfected cells) (Li et al. 2000). We further established a vaccinia virus-free system. Here, all supporting plasmids to express the N, P, and L proteins, respectively, and an additional plasmid to express T7 polymerase are driven by the cellular RNA polymerase II under the control of a CAG promoter (a combination of the cytomegalovirus early enhancer element and chicken beta-actin promoter). The plasmid to generate SeV genome RNA was left unchanged, being under the control of T7 promoter. The resulting all-plasmid-driven method demonstrated virus recovery efficiency similar to or even higher than that using UV-irradiated vTF7-3 (Nagai et al. 2007). These inventions greatly facilitated SeV engineering and vector development (see Chap. 3).

2.5 SeV Lifecycle: Revisiting Old Issues, Settling Outstanding Questions, and Raising New Issues by Reverse Genetics

SeV and most NNSVs can proliferate in enucleated cells or in the presence of DNA-dependent RNA polymerase inhibitors such as actinomycin D and α-amanitin (Pennington and Pringle 1978; Choppin and Compans 1975), indicating that their lifecycle is completed entirely in the cytoplasm without having a nuclear phase (Fig. 2.2). This characteristic is the greatest advantage of SeV-based expression vector from the aspect of safety because it principally guarantees the vector's nongenotoxic (nonintegrating) nature (see Chap. 3). This section outlines the lifecycle of SeV with special emphasis on the contribution of recombinant SeVs (rSeVs) generated by reverse genetics to its understanding.

2.5.1 Initiation of Infection

2.5.1.1 Attachment to the Target Cells

The SeV attachment protein is the HN glycoprotein. Sialo-oligosaccharides of both glycoproteins and glycolipids serve as the host cell receptors (Suzuki et al. 1983, 1985).

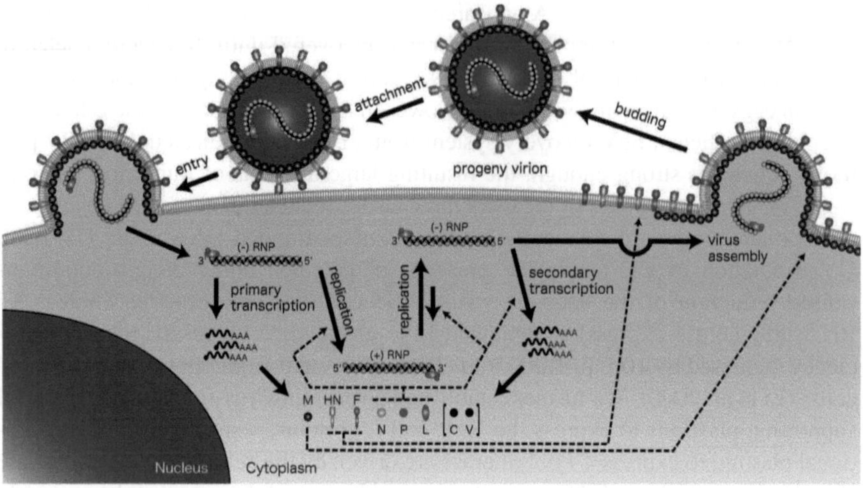

Fig. 2.2 The lifecycle of SeV (courtesy of T. Irie). For details, see text

The specificity of SeV toward receptor recognition was examined in a bovine cell line. The cells were first treated with *Vibrio cholerae* neuraminidase to remove cell-surface sialic acids to render them insusceptible to the viral infection, then modified with sialyltransferases to contain sialyloligosaccharides of defined sequences and examined for restoration of susceptibility to SeV (Markwell and Paulson 1980). Sialylation by incubation of the cells with CMP-sialic acid and β-galactoside α2,3-sialyltransferase fully restored susceptibility to infection, whereas no restoration occurred when sialylated by β-galactoside α2,6-sialyltransferase. This finding suggested preferential use of sialic acid bound to galactose by the α2,3-linkage (SAα2,3-Gal). However, SeV is usually passaged in the amnion or allantois of the chick embryo, and SAα2,3-Gal was found to be much more abundant than SAα2,6-Gal on the amniotic and allantoic cells (Sriwilaijaroen et al. 2009). Avian influenza A viruses show specificity to SAα2,3-Gal and human counterparts to SAα2,6-Gal. Passage of human viruses in the amnion or allantois resulted in a shift of receptor specificity from SAα2,6- to SAα2,3-Gal (Ito et al. 1997). Thus, SeV receptor specificity has to be revisited by using fresh virus isolates from mouse or rat (Nagai et al. 2011).

In a special case, SeV attachment, followed by entry, was suggested to occur by means of the asialoglycoprotein receptor using an SeV temperature-sensitive (ts) mutant in the HN protein (Markwell et al. 1985). This mutant (HN^-) grown at a non-permissive temperature (38 °C) lacked HA and NA activity and was unable to infect or agglutinate conventional host cells, but did attach to and infect Hep G2 cells that are known to express asialoglycoprotein receptor (ASGP-R). The ASGP-R is a special type of mammalian lectin that recognizes galactose- or *N*-acetyl galactosamine-terminal proteins. That the ASGP-R permitted the HN^- virus (possessing only the F protein) infection was verified by the observations that the infection was abolished by the competitors and antibodies specific to ASGP-R. The authors speculated that the infection would be mediated through the interaction of a terminal galactose moiety of the F protein (Yoshima et al. 1981) with ASGP-R. However, it is unlikely that galactoses become exposed as the terminal moiety for this particular ts mutant that lacks NA activity and hence is unable to remove the terminal sialic acids from the complex-type *N*-glycans of F protein (reviewed in Nagai et al. 2011). On the other hand, a more recent study demonstrated that the wild-type SeV efficiently infects cells via the asialoglycoprotein receptor and that this alternative route of cell entry also requires the fusion-competent, cleaved F glycoprotein (Bitzer et al. 1997).

2.5.1.2 Entry into the Target Cells

Two main pathways of entry of the viral genomic material into the cytoplasm are known for envelope viruses: one is direct fusion of the viral envelope with the plasma membrane and the other is the clathrin-mediated endocytosis followed by fusion between the viral envelope and the endosomal membrane. The latter route is dependent on a low pH whereas the former occurs at neutral pH as well as at a low pH. Thus, the entry via the endocytosis pathway is inhibited by lysosomotropic agents such as chloroquine and NH_4Cl that raise the endosomal pH while direct fusion with the plasma membrane is not. SeV and NDV entry was not inhibited by those agents (pH independent) under the conditions where infections by influenza A virus and VSV, which are known to use the endocytosis route, were inhibited (Nagai et al. 1983a). Morphological studies revealed fusion of the SeV envelope with the plasma membrane of the endodermal allantoic cells of chick embryo (Morgan and Howe 1968) and human erythrocyte membranes (Howe and Morgan 1969). In addition, SeV has been used as a reagent to induce cell–cell fusion and make hybrid cells (reviewed in Nagai et al. 2011). SeV and paramyxovirus entry, thus, has been believed to occur through direct fusion at the plasma membrane (Fig. 2.2). However, those data obtained by classical methods may be too simplistic to draw any conclusions. Indeed, more recent studies on RSV, a member of the subfamily *Pneumovirinae* in the *Paramyxoviridae*, showed that the infection was inhibited by (1) small interfering (si) RNAs targeting genes associated with clathrin-mediated endocytosis, (2) the expression of well-characterized dominant-negative mutants of genes in the endocytosis pathway, and (3) the clathrin endocytosis inhibitor chlorpromazine (Kolokoltsov et al. 2007). Human immunodeficiency virus (HIV) entry is also pH independent and has been thought to occur via direct fusion at the plasma membrane. However, recent live cell imaging that enabled tracking single HIV particles demonstrated that the endocytosis route is the major pathway (Miyauchi et al. 2009). SeV and paramyxovirus entry has to be reexamined using more refined experimental procedures (Nagai et al. 2011).

The paramyxovirus F glycoproteins existing as a homotrimer on the virion surface undergo dynamic conformation changes from the prefusion state to the postfusion conformation (Lamb and Parks 2007). The F protein in the prefusion state consists of a large globular head domain attached to a three-helix coiled-coil formed by HRB followed by the transmembrane (TM) domain. The HRB stalks then melt (are separated), and the HRA within the head of the prefusion state forms a coiled-coil, extends, and inserts the fusion peptide into the target cell membrane (pre-hairpin intermediate conformation). Finally, the HRBs are positioned over the coiled-coil HRA domain in an antiparallel fashion, resulting in the formation of the final post-fusion six-helix bundle (hairpin). This final step enables juxtaposing the fusion peptide-containing target cell membrane and the TM domain-containing viral membrane so that they fuse, ultimately introducing the viral genomic material into the cytoplasm (reviewed in Lamb and Parks 2007). It is believed that a virus species-specific interaction between the HN and F proteins is required for or facilitates the fusion process. Upon binding the receptor, the HN protein may undergo a conformational change that in turn could trigger the conformational change of the F protein (from the pre-fusion state to pre-hairpin intermediate conformation) (Lamb and Parks 2007).

Luque and Russell (2007) showed by plasmid-based expression studies that the point mutations introduced into the HRA either increased or decreased SeV fusogenic activity; this was now extended to the cellular level proliferation and *in vivo* pathogenicity. Namely, two rSeVs were created that possessed a point mutation, Leu179Val or Lys180Gln in the HRA, and were examined for their replication and pathogenicity. The former was found to exhibit higher cell-to-cell membrane fusion activity and increased pathogenicity for mice, compared with the parental wild type, whereas the latter exhibited reduced levels of F-protein expression, cleavage, and fusogenicity at the cellular level and pathogenicity *in vivo* (Luque et al. 2010).

2.5.2 *Gene Expression*

2.5.2.1 Requirement of *De Novo* Synthesized N and P Proteins in the Primary Transcription

After entering the cytoplasm, the SeV and paramyxovirus genomes do not become naked but remain encapsidated. This encapsidated negative strand RNA genome, namely the (–)RNP, is the template for both transcription and replication (Fig. 2.2). The transcription is initiated by the virion-associated, virus-specific RNA polymerase comprising the L and P proteins (Hamaguchi et al. 1983; Horikami et al. 1992). This process is termed the primary transcription (Fig. 2.2). The primary transcription is thus a phase of viral gene expression catalyzed solely by the polymerase associated with infecting RNP without the support of any *de novo* synthesized viral proteins. The secondary transcription occurs using *de novo* replicated (–)RNP template and the *de novo* synthesized viral proteins (Fig. 2.2) (see Sect. 2.5.3.1).

Analysis of the primary transcription process has often been done in a test tube using isolated RNPs from the virions and L–P RNA polymerase and cellular factors recovered from infected cells, but this setup requires extremely large copy numbers of both the template and the proteins. At the infected cellular level, it has been virtually the only way to see what happens in the presence of cycloheximide. Here, again a large input [extremely high multiplicity of infection (moi)] is necessary. These approaches thus would be quite different from a natural course of infection, which is principally initiated with one infectious unit/cell. To overcome this problem, it would be necessary to dissect the primary transcription phase from the complex natural course of infection. Wiegand et al. (2007) created rSeV deletion (Δ) mutants that infect cells but cannot undergo replication and encapsidation and therefore cannot advance to the phase of secondary transcription: these were Δ*N* with the entire N ORF deleted and Δ*P* with the entire P ORF deleted. The primary transcription of Δ*N* and Δ*P* in the absence of *de novo* viral protein synthesis occurred at a hardly detectable level. In contrast, when newly synthesized N proteins were supplied in *trans*, the primary transcription reached remarkably higher (likely normal) levels. In the subsequent study, an additional mutant *P*Δ2–77 was created, which lacked the N-terminal 71 residues of the P protein required for binding the newly synthesized N (N^0) proteins and hence could not support the nascent RNA chain assembly with the N^0 proteins to generate (+)RNPs (see Sects. 2.5.3.1 and 2.5.4.1). It was found that although the primary transcription occurred at a low level in infection with Δ*P*, the primary transcription levels rose to eightfold in *P*Δ2–77 infection (Bossow et al. 2012). It was concluded that *de novo* synthesis of N and P proteins could be a key event that is required for the transition from the preliminary (early) to normal (late) primary transcriptional phase so that the lifecycle can enter the replication phase followed by the secondary transcription (for details, see Chap. 4). Curran and Kolakofsky (2008) also presented evidence suggesting that SeV soluble N is not only required for replication (see Sect. 2.5.3.1) but also stimulates transcription.

2.5.2.2 Nonlinear Attenuation

Vesicular stomatitis virus VSV in the *Rhabodoviridae* is a prototypic NNSV. Our view on the transcription of paramyxoviruses has been built largely depending on the studies on the transcription of VSV. In common, there are *cis*-acting elements to terminate (stop) the upstream gene transcription and to reinitiate (start) the downstream next gene transcription (see Sect. 2.4.1). The entry of the polymerase (L–P complex) at the very 3′-end of the genome and polar, sequential transcription of each gene by the stop-and-restart mechanism toward the 5′-end (the VSV paradigm of transcription) has been applied to SeV and other NNSVs. However, this polar transcription does not generate each mRNA in equimolar quantities in infected cells, perhaps because the reinitiation efficiency of transcription at each gene boundary is high, but not perfect. Thus, it has been believed that there is linear attenuation of transcription toward the 5′-end, where the *N*-gene product as well as the leader transcript (le+) is the most abundant and the *L*-gene product is the least abundant (Whelan and Wertz 2002; reviewed in Whelan et al. 2004). However, the entry not only at the very 3′-end but also to the *N*-gene start site was also proposed for VSV polymerase. Although the 3′-end entry could be relevant for both transcription and replication, the *N*-gene start entry could be operative for only transcription (see Sect. 2.5.3.1).

For SeV, the *P*, *M*, and *HN* gene starts are 3′-UCCCACUUUC-5′, whereas the *N* start is 3′-UCCCAgUUUC-5′, the *F* start 3′-UCCCuaUUUC-5′, and the *L* start 3′-UCCCCACUUaC-5′. Using rSeVs expressing a reporter gene under the control of different start signals, the *F*-gene start was found to direct a lower level (about one fifth) of transcription, compared with the other three whose initiation capacities were almost equivalently high (Kato et al. 1999). Thus, the expression of the *F* gene and the downstream *HN* and *L* genes was remarkably downregulated in the natural lifecycle of the parental SeV in cells, whereas the expression levels of the first three genes (*N*, *P*, and *M*) were nearly equivalent. The rSeV in which the authentic *F* gene start of the wild type was converted to that shared with the *P*, *M*, and *HN* genes expressed significantly more F proteins as well as the two downstream gene products, HN and L, replicated with faster kinetics in cultured cells and eggs, and was more virulent for mice, exhibiting a 20-fold-lower 50 % lethal dose (LD_{50}) (Kato et al. 1999). Therefore, SeV transcription appears to undergo not a linear, but rather a biphasic attenuation with a remarkable down-modulation between the *M* and *F* genes. Although the *HN* and *L* starts themselves possessed high reinitiation capacity, expression of these two genes remained as low as *F*-gene expression in the wild-type infection because of that moderation of *F*-gene expression, indicating no polymerase entry at the *HN* and *L* start signals. This observation meets the 3′- to 5′-polar transcription model. Identical sequence differences in the start signals are seen in all SeV isolates sequenced to date, suggesting that the variations are locus specific (Kato et al. 1999). The gene start signal-mediated moderation of F expression thus appears to be universal for SeV, suggesting its selective advantage for the virus in nature. The nonconserved positions within the conserved *cis*-acting elements appear to be highly relevant to the biology of SeV in the natural settings (Nagai et al. 2011).

2.5.3 Genome Replication

2.5.3.1 Overview

After accumulation of viral proteins translated from the primary transcripts, the 15,384-nt-long genome (–) RNA is replicated to a full-length exact copy, the antigenomic (+) RNA (Fig. 2.2). The same L–P complex as used for transcription is believed to act as the replicase, but the former is finely tuned by the transcription regulatory signals (see Sect. 2.5.2.2), whereas the latter must ignore these signals and read through the intergenic sequence. Thus, the mechanism of switching from transcription to replication has been a central issue in NNSV biology. One of the key factors of those involved in this switching is the concomitant, continuous association of *de novo* synthesized soluble N proteins with the nascent, growing chain of RNA in the direction from the 5′- to 3′-end (Gubbay et al. 2001). The P protein is essential for maintaining the N protein in a soluble form (N^0) and chaperones its association with the nascent viral RNA chain without allowing its illegitimate association with nonviral RNAs (Curran et al. 1993, 1994).

For VSV RNA polymerase, a two-entry site model has also been proposed as noted above. In this model, the polymerase is thought to initiate synthesis of the le+ and the full-length antigenome at the very 3′-end (position 1); however, for mRNA synthesis, the polymerase initiates by entry directly at the first gene start site (Whelan and Wertz 2002; reviewed in Whelan et al. 2004). Alternatively, the polymerase was thought to enter the genome at the very 3′-end, but for transcription scans through the leader region to initiate mRNA synthesis at the first gene start site. Qanungo et al. (2004) isolated two different kinds of VSV RNA polymerase complex from infected cells. One complex was the transcriptase that contained host proteins EF1 (translation elongation factor 1α) and hsp 60 (heat-shock protein 60), and started RNA synthesis exclusively at the first *N*-gene start site. The other was the replicase that contained soluble N but not EF1 or hsp 60 and started RNA synthesis exclusively at the genome 3′-end. Soluble N thus appears to be essential for the polymerase to enter the genome 3′-end to initiate replication. This multiple- (or two-) site entry model is also applicable to transcription (see Sect. 2.5.2.2). These issues have to be addressed with SeV and other paramyxoviruses.

As viral protein levels increase, the (+)RNP ⇆ (−)RNP replication cycle is facilitated (Fig. 2.2). Here, the (−)RNP acts as the template for the secondary transcription to raise the viral protein level on one hand and on the other is packaged in the progeny virion (Fig. 2.2). This synchronized active protein synthesis and genome replication eventually lead to vigorous production of progeny virions.

2.5.3.2 The Rule of Six and Replication Promoters

Each N protein of SeV recognizes six nucleotides (hexamer) (Egelman et al. 1989), and 2,564 monomers of N protein are required to finish the assembled 15,384-nt-long SeV genome or antigenome. Mutagenesis studies on SeV model minigenomes revealed that efficient replication occurs only when the total number of nucleotides of a genomic or antigenomic RNA is an even multiple of six (rule of six) (Calain and Roux 1993; Kolakofsky et al. 1998). There are two discontinuous replication promoter elements, PrE-1 and PrE-II, each consisting of three hexamers (Fig. 2.1c; reviewed in Lamb and Parks 2007). The 3′-terminal 12 residues (two hexamers) in PrE-1 are identical except for the fifth position between the genome and antigenome (Fig. 2.1c). PrE-II consists of (3′-CNNNNN-5′) × 3 positioning at the 14th, 15th, and 16th hexamers in both the genome and antigenome (Tapparel et al. 1998) (Fig. 2.1c). PrE-1 and PrE-II are, thus, separated by 60 nucleotides (ten hexamers). Changing the length of this spacer impaired replication, probably because the N-bound hexamers 1st to 3rd and 14th to 16th have to be juxtaposed on the same surface of coiled RNP with a turn of 13 Ns for the RNA polymerase to recognize the template and initiate replication of the genome and antigenome (Tapparel et al. 1998; reviewed in Lamb and Parks 2007) (Fig. 2.1b).

The importance of the nucleotide sequences in the PrE-II (nt 79–96) (Fig. 2.1c) and their correct positioning also appeared to be supported by an experiment trying to insert an 18-nt sequence (three hexamers) containing a unique *Not*I restriction site into the SeV genome for a foreign gene insertion. When the 18-nt sequence was inserted between nt 79 and 80 within the first hexamer of PrE-II (Fig. 2.1c), no infectious virus could be recovered, whereas when inserted far downstream of PrEII (between nt 119 and 120), the virus was readily recovered (Hasan et al. 1997). The 18-nt insertion was possible at many places as far as the correct positioning of PrE-I and PrE-II in the genome and antigenome was not disturbed (see Chap. 3).

2.5.3.3 *Trans*-Regulatory Mechanism for Determining the Negative Polarity of SeV Genome. The SeV Paradigm vs. VSV Paradigm

How an NNSV can be an NNSV predominantly possessing (–) RNP is a central question in virology. It is believed that there is no mechanism to selectively incorporate only the (–) RNP into the virions during the process of assembly. Indeed, for SeV, genomic and antigenomic RNA containing SeV virions were shown to be produced in proportion to their intracellular ratio, although antigenome-containing virions were not infectious (Kolakofsky et al. 1974; Kolakofsky and Bruschi 1975; Mottet and Roux 1989).

Using the prototypic species, VSV, it has been established that the negative polarity of the viral genome is defined solely by *cis*-acting replication promoters; the promoter of replication-intermediate positive sense antigenome RNA [the sequence complementary to the 5′-terminal trailer (*tr*) sequence of the genome] to direct negative sense genome RNA was much stronger than that of the promoter, 3′-leader sequence (*le*) of the genome to direct positive sense antigenome RNA, thus resulting in the preferential production of progeny with the negative strand RNA genome. Accordingly, many more genomes than antigenomes are synthesized in cells throughout the single-cycle replication of VSV (Lyles and Rupprecht 2007).

The *cis*-acting promoter-dependent asymmetry concept (the VSV paradigm) appears to have been applied to other NNSVs including SeV without concrete results. The application of the VSV paradigm appears to be justified, for example, by the following experimental results and the interpretation that came from studies using SeV internal-deletion and copy-back DI (defective interfering) particles. The internal-deletion DI particle genomes contain the authentic *le* and *tr* sequences present in the full-length (–) RNA, whereas the *le* sequence in the copy-back DI RNAs is the exact copy of the 5′-terminal *tr* sequence of the genome. Thus, the 3′-terminal sequences of PrE-I (Fig. 2.1c) differ between the internal-deletion DI genome and copy-back DI genome. Both of the DI types contained the same PrE-II. Mixed infections of eggs with SeV stocks containing internal-deletion DI and copy-back DI genomes followed by serial egg passages outcompeted internally deletion-type DI particle genomes, resulting in predominance of copy-back DI genomes and extinction of internal-deletion DI genomes (Re and Kingsbury 1986; reviewed in Re 1991). This out-competition was not caused by size differences between the internal-deletion DI and copy-back DI genomes nor whether the transcription was inert (copy-back DI) or active (internal-deletion DI). Thus, the interpretation was that the

3′-end promoter in the antigenome could be stronger than the 3′-end promoter in the genomic RNA, meeting the VSV paradigm (Lamb and Parks 2007).

However, the interpretation cannot be that simple because the analyses were carried out under conditions where the C proteins were expressed from the helper virus to replicate the DI genomes and because the C proteins were found to have a capacity to down-modulate SeV mRNA synthesis (Curran et al. 1992) and RNA replication selectively via the genomic promoter (Cadd et al. 1996; Tapparel et al. 1997; Horikami et al. 1997).

Late in the natural lifecycle of SeV, eight to ten times as many genomes as antigenomes are produced. However, it was found that the quantitative ratios of genome to antigenome in cells was not constant during the single-cycle replication of SeV and that a marked shift from the early antigenome-dominant to the late genome-dominant phase took place, in contrast to the case of VSV in which no such shift occurred and many more genomes than antigenomes were always synthesized (Irie et al. 2014).

The shift of SeV genome polarity in cells appeared to be governed primarily by the expression of the accessory C protein because no such shift occurred in cells infected with 4C(–) rSeV expressing none of the four C accessory proteins (see Sect. 2.5.5.3); antigenomes were dominant throughout infection, generating antigenome-dominant virions (Irie et al. 2008a, 2014). Thus, the infectivity to HAU (reflecting the physical particle numbers) ratio was 1 to 2 logs lower compared with that of the wild type (Hasan et al. 2000). The antigenome dominance in the absence of RNA-inhibiting C proteins indicated that the genomic promoter is intrinsically stronger than the antigenomic promoter.

Overall, SeV genome polarity determination did not appear to meet the VSV paradigm. It appeared to be governed by a *trans*-regulatory mechanism, raising a novel SeV paradigm. The early antigenome-dominant phase involved increased synthesis of mRNAs that facilitates the amplification of viral proteins, including the C proteins for the antigenome to genome switching.

It is surprising that just the accessory gene products, which although widespread, are not ubiquitous in paramyxoviruses, play such an important role in leading to predominance of genomic RNA, thereby rendering SeV an NNSV. It is intriguing to learn which of the two paradigms is applied to other NNSVs. There may be an yet unknown third paradigm.

2.5.4 Interacting and Functional Domains and Motifs in the P, N, and L Proteins

Extensive studies have been made on the interactions between the involved players to gain deeper insight into the paramyxovirus RNA synthesis machinery (reviewed in Lamb and Parks 2007). Here, we try to summarize briefly the interactions between the P, N, L, and genomic RNA specifically focusing on SeV (reviewed in Nagai et al. 2011).

2.5.4.1 The P Protein

The SeV P protein (568 aa) can be divided into two segments, the N-terminal PNT (residues 1–319) and the C-terminal PCT (residues 320–568) (Blanchard et al. 2004). The PNT is unstructured and contains the site (residues 33–41) to bind newly synthesized, soluble N^0 protein, chaperoning its association with the nascent RNA chain during RNP elongation (Curran et al. 1995) (see Sect. 2.5.3.1). All other known activities have been mapped to the PCT: these include the homo-tetramer formation domain (PMD) (residues 320–433) (Tarbouriech et al. 2000a, b), the L-binding domain (residues 412–445) (Curran et al. 1994; Smallwood et al. 1994), which is in part within the PMD, and a predicted triple α-helical bundle of the C terminus (residues 479–568) that is suggested to bind the N protein of the RNP template (Ryan et al. 1991; Curran 1998). This module of RNP-binding domain is expressed as the X protein in the natural lifecycle through scanning-independent initiation of the P mRNA (Curran and Kolakofsky 1988b). When expressed alone in *Escherichia coli*, the X protein is extremely soluble and monomeric (Tarbourriech et al. 2000a). Nuclear magnetic resonance (NMR) revealed that the X protein consisted of N- and C-terminal subdomains linked by an 11-residue-long linker and that the latter domain contained three α-helices (α1, α2, and α3) (Blanchard et al. 2004). The surface created by the α2 and α3 of X binds a region in the tail of the N protein (see Sect. 2.5.4.2), which also has a helical propensity, with a 1:1 stoichiometry (Houben et al. 2007). The C-terminal ~40 % (residues 344–568) is competent for mRNA synthesis *in vitro* but incapable of replication (Curran 1996).

2.5.4.2 The N Protein

The N protein (524 aa) is also divided into two segments, the N-terminal N core (residues 1–399) and the C-terminal N tail (residues 400–524) (Buchholz et al. 1993). An N-terminal domain (residues 114–129) is required for RNP to function as a template in viral replication (Myers and Moyer 1997). The N core is involved in N–N and N–RNA interactions driving RNP assembly and RNA replication (Buchholz et al. 1993). The central region of N core (residues 258–369), which is highly conserved for paramyxoviruses and contains the 10-residue-long motif (residues 262–271), could be essential for self-assembly of N with RNA. The N-terminal 255 residues are also thought to be essential for self-assembly of N with RNA (Myers and Moyer 1997). The N tail is also required for template function (Curran et al. 1993). As already noted, it contains the site to bind the P (in the L–P complex) (residues 462–471) (Çevik et al. 2004), which is followed by the two predicted α-helices (residues 477–493 and 513–520, respectively) (Houben et al. 2007). Treatment of SeV RNP with trypsin removed a C-terminal portion of the N tail: this yielded a more rigid and yet RNase-insensitive structure (Heggeness et al. 1981), suggesting that the Ntail confers flexibility of the coiling of the native RNP and its functional diversity.

2.5.4.3 The L Protein

The SeV L protein (2228 aa) is believed to possess all the enzymatic activities required for RNA synthesis, including nucleotide polymerization as well as 5′-end capping and methylation of mRNAs (Ogino et al. 2005; reviewed in Lamb and Parks 2007). Polyadenylation of mRNAs occurs through a mechanism of polymerase stuttering at the 5 Us stretch in the end of transcription stop signals (see Sect 2.4.1). These functions are executed with association of the P protein. Six domains (I–VI) are conserved for the L proteins of NNSV (Poch et al. 1990). For SeV they are domains I (residues 225–416), II (503–607), III (653–876), IV (927–1128), V (1129–1378) and VI (1770–1847) (Feller et al. 2000).

A series of extensive mutagenesis including point mutations, deletions, insertions, and swapping with the corresponding measles virus residues within each of the six domains indicated that the activities of SeV L protein are not simply compartmentalized in a particular domain, but rather that each domain contributes to multiple steps in viral RNA synthesis (Chandrika et al. 1995; Feller et al. 2000; Cortese et al. 2000; Smallwood et al. 2002a). Thus, pairs of mutated inactive L proteins could sometimes restore viral RNA synthesis. This type of complementation occurred with mutants in five of the six domains, the only exception being domain III (Smallwood et al. 2002b). However, a catalytic residue for mRNA cap formation was pinpointed at the Lys1782 in domain VI, and the rSeV with the Lys1782–Ala change completely abolished cap methylation with greatly attenuated growth kinetics in Vero cells (Murphy and Grdzelishvili 2009). It is interesting to note that the transcription and replication were uncoupled for considerably many mutants.

The RNA synthesis inhibition and the inhibition of complex formation with the P protein are largely independent because many of the L mutants incapable of RNA synthesis were capable of P binding. The P-binding site was mapped to residues 1–350 (Holmes and Moyer 2002). The N-terminal 178 residues preceding the domain I were required for homo-oligomerization. Two regions (residues 2–19 and 20–178, respectively) appeared to independently mediate homo-oligomerization because deletion of either region still supported oligomerization, although this abolished RNA synthesis (Çevik et al. 2007). The interaction of L protein and C protein could be relevant to the promoter-dependent RNA synthesis inhibition by the latter (see Sect. 2.5.3.3). The region in the L protein involved in this interaction was mapped to the N-terminal 895 residues containing domains I to III (Horikami et al. 1997), and that in the C protein to the C-terminal half (Grogan and Moyer 2001).

2.5.5 *Dispensability of the Accessory V and C Proteins for the Cellular Level Proliferation*

2.5.5.1 The Expression Strategies of the *P/C/V* Gene

The paramyxovirus V and C proteins were first identified for SeV (Lamb and Choppin 1977a). They were originally considered to be nonstructural proteins

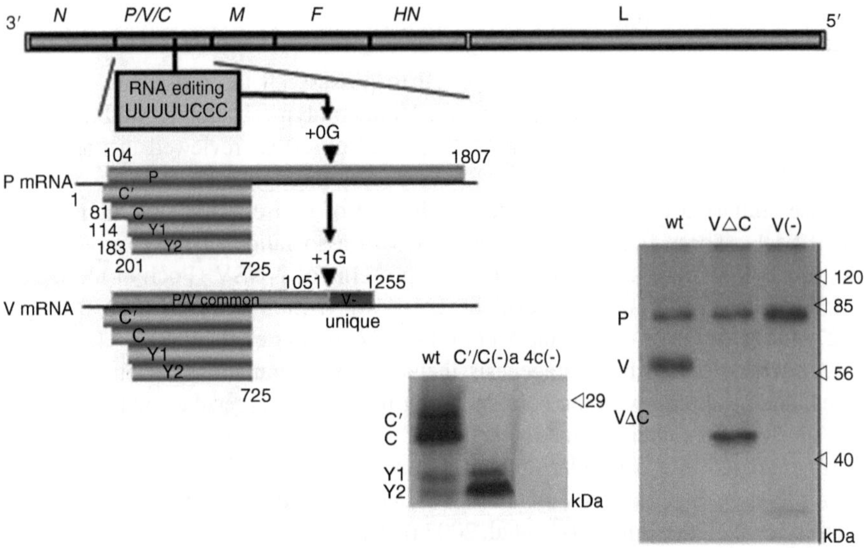

Fig. 2.3 Creation of V(−), VΔC, C′/C(−), and 4C(−) viruses by SeV reverse genetics. For details, see text. (Adapted from Nagai et al. 2011, with permission)

expressed in infected cells, but not present in the virions. Later, however, the presence of C protein in a small, but appreciable amount in both the SeV virion and RNP was demonstrated (Portner et al. 1986; Yamada et al. 1990). The V and C proteins are not ubiquitous, although widespread, in the *Paramyxovirinae* and are hence called accessory proteins.

The SeV V protein is produced by a transcriptional frameshift (co-transcriptional RNA editing). The P mRNA is the faithful copy of the *P/C/V* gene, accounting for about three fourths of the total transcripts in cells and giving rise to the P protein (568 aa), the smaller subunit of the RNA polymerase complex. The remaining transcripts (one fourth) are the V mRNA, which carry a single guanine residue (+1G) (at nt position 1052) inserted by polymerase stuttering at a specific editing site (3′-UUUUUCCC-5′) located midway down the *P* gene; this insertion shifts the reading frame to fuse the upstream P/V common ORF encoding 316 residues to the downstream -1 ORF of V-unique (Vu) 68 residues (Vidal et al. 1990a,b; Curran et al. 1991) (Fig. 2.3). The V protein, therefore, consists of the P/V common region (the N-terminal approximately four fifths before the frameshift) and the Vu region (the C-terminal approximately one fifth after the frameshift). Two Gs insertion can also occur to yield the W protein that terminates three codons downstream. This transcript accounted for less than 5 % of the total (Kato et al. 1997b) and is not addressed here.

All five genera of the subfamily *Paramyxovirinae* (*Respirovirus*, *Morbillivirus*, *Henipavirus*, *Rubulavirus*, and *Avulavirus*) express the V protein from the respective *P* genes with the exception of human parainfluenza virus (HPIV) 1 and HPIV3 in the *Respirovirus*. The Vu region of paramyxoviruses exhibits a perfect preservation of seven cysteine residues that form a zinc finger-like motif and indeed bind two atoms of Zn^{2+} for at least some members including SeV (Fukuhara et al. 2002). HPIV1 expresses no V protein because it does not undergo editing (Matsuoka et al. 1991). However, there is a characteristic, cysteine-rich V relic for HPIV1. The *trans* V frame of HPIV3 is open but is separated from the editing site by several stop codons (Galinski et al. 1992). These results suggested that both HPIV1 and HPIV3 once expressed a V-like protein (Nagai 1999; Nagai and Kato 1999).

An open reading frame (ORF), which overlaps the N-terminal P (P/V common) ORF in the +1 frame, gives rise to four proteins: C′ (215 aa), C (204 aa), Y1 (181 aa), and Y2 (175 aa), collectively referred to as the C proteins, that initiate at a non-AUG codon, ACG/81 and AUGs/114, 183, 201, respectively, and terminate at 725 (Curran and Kolakofsky 1988a, 1989; Gupta and Patwardhan 1988) (Fig. 2.3). Although the C′ and C proteins are translated by a leaky scanning mechanism, the Y1 and Y2 are translated through a scanning-independent mechanism (Curran and Kolakofsky 1988a, 1989) or, more precisely, via discontinuous scanning (ribosomal shunt) (Latorre et al. 1998). The Y1 and Y2 were suggested to occur not only by *de novo* translation initiation but also by proteolytic processing of the C′ (de Breyne et al. 2004). Among the four C proteins, C is the major species expressed in infected cells at a molar ratio several fold higher than those of the other three (Curran and Kolakofsky 1989; Kurotani et al. 1998). The three genera *Respirovirus*, *Morbillivirus*, and *Henipavirus* express the C protein. Its homology is remarkably high within each genus, but quite poor between the different genera; in common, however, they are relatively small (~200 residues) and highly basic with the pIs around 10 (Nagai 1999; Nagai and Kato 2004).

One additional protein called X is synthesized from the P frame that is encoded more than 1,500 nt downstream of the 5′-end of P mRNA and expressed by scanning-independent ribosomal initiation (Curran and Kolakofsky 1987, 1988b) (see Sect. 2.5.4.1).

2.5.5.2 A Strategy for Opening Two Overlapping ORFs in an Error-Prone, Single-Stranded RNA Genome

Because of the lack of proof-reading and repair mechanisms, RNA virus genomes are error prone. It is thus even surprising that SeV *P* gene of only 1,893 nt opens overlapping frames to express three different protein species: P, V, and C.

Nonsynonymous nucleotide substitutions (Ka) are caused by various selection pressures, whereas synonymous substitutions (Ks) are largely neutral, free from selection pressures, and accumulate as the rounds of replication increase. Ks occur at a nearly constant rate in the variable and conserved regions for paramyxovirus monocistronic genes (Sakaguchi et al. 1989). In fact, Ks values were within a relatively narrow range (0.351–0.450) for the five (N, M, F, HN, and L) monocistronic ORFs of SeV, and Ka/Ks values also converged to a narrow range (0.029–0.069), indicating that neutral changes have accumulated at a nearly constant rate over these different ORFs (Fujii et al. 2001; Nagai et al. 2011). On the other hand, when P/C overlapping frames were divided into two segments (the N-terminal one third and C-terminal two thirds) and analyzed, it was found that the SeV *P/C* gene might have evolved by positioning a conserved subdomain ORF in parallel with a variable ORF. For example, in the C-terminal two thirds, the Ks value for C ORF (residues 73–204) was within a normal range (0.388), while its Ka was extremely low (0.003), giving rise to an extremely low Ka/Ks value (0.008). In contrast, for the counterpart P ORF (residues 76–208) Ks value was unusually low (0.046), Ka considerably high (0.105), and as a result, Ka/Ks unusually high (2.283). These data indicated that the C ORF in that subsegment had been subject to little change and that the accumulated synonymous changes had been dictated by the counterpart P ORF as nonsynonymous but yet tolerable changes. In other words, the P ORF of the subdomain was flexible enough to buffer the synonymous nucleotide changes accumulating in the C ORF (Nagai et al. 2011). Such high variability of the P ORF segment and low variability of C ORF segment could be relevant to the fact that no particular functional and interacting domains have been mapped so far within the P ORF segment (see Sect. 2.5.4.1), whereas major functions such as interferon antagonism and RNA synthesis inhibition of the C protein have been mapped to its C-terminal half within that C ORF segment (see Sect. 2.7.1). The N-terminal one-third overlapping ORFs (residues 5–75 for the P and 1–72 for the C) appeared to have evolved in a similar way; however, the C ORF segment acted as a buffer to maintain relative conservation of the P ORF segment (Nagai et al. 2011). This P ORF segment contains a site essential for soluble N binding and RNP elongation (see Sects. 2.5.3.1 and 2.5.4.1). Because of the short length, statistically reliable interpretation was difficult for the Vu and its P ORF counterpart.

2.5.5.3 Generation of V and C Knockout Viruses and Their Phenotypes at the Cellular Level

One of the long-held major questions was whether the SeV V and C proteins meet the criterion of the accessory gene products; they must be able to be deleted without fully destroying the viral infectivity at the cellular level. Without affecting the P ORF, changes in two nucleotides (underlined) were introduced into the editing site (UCUUUCUC) (Fig. 2.3) or a stop codon was introduced just downstream of the editing site. Mutants with the total V protein deleted [V(−)] and only the C-terminal V-unique region deleted (VΔC) were respectively recovered without any difficulty just after propagation of transfected cells in eggs (Kato et al. 1997a, b) (Fig. 2.3). Essentially the same deletion mutant as VΔC was created by Delenda et al. (1997). These mutants replicated in cells as efficiently as the wild-type SeV and their ratios of infectivity (PFU/ml) to the physical particle numbers [substituted by the hemagglutination units (HAU)] were as high as that of the wild type recovered from the cDNA, establishing that the V protein is nonessential and that both of the V mutants are fully infectious. The role of V protein, if any, in the lifecycle therefore remains to be elucidated. It was found that SeV gene expression and genome replication were considerably augmented in the absence of entire V protein (i.e., in infection with V(−) but not with VΔC), suggesting some moderating role of V protein in SeV transcription and/or replication. Because of such augmenting nature, SeV vector in the V(−) virus backbone has sometimes been used to enhance the *trans*-gene expression at the cellular level (see Chap. 3). However, V(−) displayed higher cytopathogenicity than the wild type or VΔC (Nagai and Kato 2004; Nagai et al. 2011).

Through a difficult process including at least three passages in eggs, a viable clone, 4C(−), which expressed none of the four C proteins (C′, C, Y1, and Y2), was recovered (Fig. 2.3) by putting stop codons just after the initiation codons of the respective ORFs, again without affecting the P ORF (Kurotani et al. 1998). The difficulty of recovery was the result of a reduction of 4C(−) virus infectivity by about 5 logs, and the success of recovery probably was attributable to the high efficiency of our SeV rescue system (see Sect. 2.4.3). Four successive egg passages were required to obtain detectable HAU for 4C(−) (Nagai et al. 2011).

Studies on the cellular phenotypes of 4C(−) indicated that the C proteins determine the genome polarity by suppressing genome promoter-driven RNA synthesis (see Sect. 2.5.3.3), facilitates the virus assembly and budding (see Sect. 2.5.7.3), and encodes antiapoptosis function (Koyama et al. 2001, 2003). The C proteins were further found to inhibit IFN-α/β production and counteract the antiviral action by IFN-α/β and IFN-γ (see Sect. 2.7.1). These results indicated that the SeV C proteins also fall in a category of nonessential accessory gene products but are extremely versatile.

Three clones, C′/C(−)a, C′/C(−)b, and C′/C(−)c, that expressed Y1 and Y2 but not C′ and C, were isolated with relative ease, compared with 4C(−). Their infectivities were, however, 2 logs lower than that of the wild type and their PFU to HAU ratios were only about one tenth of the wild type. Although C′ and C expression was indeed successfully eliminated, the levels of Y1 and Y2 expression were remarkably increased (Kurotani et al. 1998) [representatively shown with C′/C(−)a in Fig. 2.3]. This result suggests that the increased Y1 and Y2 expression compensated for at least part of the functions commonly shared with the C protein, the major species of the four C proteins, and the smaller Y1 and Y2 (see Sect. 2.5.5.1). The result further would provide a remarkable case in which translation at an internal AUG codon in mammalian cells can be facilitated by the introduction of a stop codon within the upstream ORF (in this case the C′ and C ORFs) (Hughes et al. 1984; Liu et al. 1984; Kozak 1984; Kurotani et al. 1998). Facilitation was more significant for Y2 than for Y1, in agreement with the notion that the efficiency of translation reinitiation steadily improves as the distance from the upstream ORF increases (Kozak 1987). All the phenotypes displayed by the three C′/C(−) clones in cell culture and in mice have so far been essentially identical (Nagai et al. 2011).

2.5.6 *Posttranslational Modifications*

2.5.6.1 Phosphorylation

Because of heavy phosphorylation, the paramyxovirus phosphoprotein (P) is so named. The primary phosphorylation site is the serine residue at 249 in the SeV P protein. This phosphorylation is thought to be carried out by protein kinase Cζ

(Huntley et al. 1997). The SeV L protein may also be involved as it was found to phosphorylate the P protein as well as the N protein *in vitro* (Einberger et al. 1990). Ser249 phosphorylation-deficient rSeVs were generated by introducing a mutation Ser249Ala, Ser249Asp, or Pro250Ala. These mutants were comparable to the wild type with respect to replication in cell cultures and pathogenicity for mice as well as oligomerization (Hu et al. 1999). Because Ser249 is localized within the P/V common region, V protein phosphorylation at Ser249 was also abrogated in these mutants. Thus, primary phosphorylation at residue 249 of the P and V proteins appeared to have no essential role in SeV transcription and replication. However, the situation is complex because alternate serine residues have been phosphorylated in these mutants. These alternate sites were further knocked out so that the net phosphorylation was reduced to about 10 % of the wild type. These mutants were found to be fully competent at least in supporting minigenome replication (Hu and Gupta 2000), whereas their behavior during the natural lifecycle remains to be clarified (Nagai et al. 2011).

At least a portion of intracellular M proteins was found to be phosphorylated whereas those in the virions were not (Lamb and Choppin 1977b). However, this phosphorylation appeared to be just an epiphenomenon irrelevant to either the lifecycle or pathogenesis (see Sect. 2.5.7.2). The N-terminal regions of C′ and C extending from the Y1 have a serine residue phosphorylated (Hendricks et al. 1993). In infected cells, they are present in phosphorylated and unphosphorylated forms. The SeV N and HN proteins are also phosphorylated in cells (Einberger et al. 1990; Takimoto et al. 1998). The biological significance for these phosphorylation events remains to be elucidated (Nagai et al. 2011).

2.5.6.2 Glycosylation and Disulfide Bridging

The carbohydrate structures of egg-grown SeV HN and F glycoproteins were analyzed. Both glycoproteins contained fucose, mannose, galactose, and glucosamine, but not galactosamine, indicating that their sugar chains are exclusively of the asparagine-linked (*N*-linked) type (Yoshima et al. 1981). These authors further showed that two main classes of oligosaccharides, the complex type and the high-mannose type, are present in both of the glycoproteins and that the terminal moiety of the former is exclusively a galactose, but not a sialic acid, because the preexisting terminal sialic acid has been removed by the viral NA activity (Nagai et al. 2011).

The HN protein contains four potential *N*-glycosylation sites: Asn77, Asn448, Asn499, and Asn511. Site-directed mutagenesis and plasmid-based expression revealed that oligosaccharide chains are attached to three of these (Asn77, Asn499, and Asn511), but not to Asn448, and that elimination of the sites Asn499 and Asn511 by replacement of the asparagine residue with glutamine residue (Asn499Gln and Asn511Gln) resulted in the most detrimental outcomes, significantly decreasing cell-surface expression, hemadsorption, and neuraminidase activity (Segawa et al. 2003). In addition, they were unable to assist syncytia formation by the F glycoprotein coexpressed (see Sect. 2.5.1.2). By a similar approach, all the three potential *N*-glycosylation sites, Asn104, Asn245 and Asn449, of the SeV F protein were shown to be glycosylated. Using an anti-F monoclonal antibody specific for native conformation and cell fusion assay with HN coexpression, deglycosylation at the site Asn245 was found to critically impair efficient intracellular transport of the F protein, its cell-surface expression, and cell-fusing capacity, whereas the mutations of the other two sites were not (Segawa et al. 2000). The *N*-linked oligosaccharide chains of SeV HN and F protein thus appeared to affect their interactions with the endoplasmic molecular chaperons and their folding and intracellular transport in a position-dependent manner (Tamura et al. 2002) (reviewed in Nagai et al. 2011).

Protein molecules expressed on the surface of plasma membrane or secreted to the extracellular matrix are directly exposed to extracellular conditions. To help maintain their structures, the polypeptide chains are often stabilized by covalent crosslinkage, most often by disulfide bridging. Generally, disulfide bonds are not formed in the cytosol, where a high concentration of reducing agents converts S–S bonds back to cysteine SH groups; hence, disulfide bonding is an event executed in the endoplasmic reticulum. In fact, all the five cysteine residues in the cytosolic M protein of SeV remain in a reduced state (see Sect. 2.5.7.2). There are ten cysteine residues in the SeV F protein, which are perfectly conserved in other paramyxovirus F proteins. Assignment of disulfide bridges in the SeV F protein using cystine-containing peptide fragments isolated from highly purified F protein revealed that Cys70, the only cysteine in F_2 and Cys199, the most upstream one in F_1, form the F_1–F_2 interchain bond. The F_1 intrachain bonds are Cys338 to Cys347, Cys362 to Cys370, the most downstream Cys424 to either Cys394, Cys399, or Cys401, and between the remaining two. It thus appears that the cysteine-rich region in the middle of F1 subunit contributes to the formation of a bunched structure containing at least three tandem cysteine loops (Iwata et al. 1994). No relevant information is available for the HN protein. To gain more detailed structural insight it is essential to resolve the three-dimensional (3D) structures specific for SeV F and HN proteins.

2.5.6.3 Cleavage Activation

SeV is routinely propagated in the allantoic cavity or amniotic cavity of embryonated chicken eggs. The egg-grown SeV is highly infectious to a wide variety of cells of mammalian and avian origin in culture as well as to eggs and possesses cell fusing and erythrocyte-lysing (hemolytic) capacity. However, when the active egg-grown virus was propagated in cultured cells, the progeny virus does not display any infectivity, hemolytic activity, or fusion activity. It was established that cell culture-grown SeV possesses the biologically inactive precursor F_0 protein. The F_0 is cleaved at a specific site (between Arg116 and Phe117) by trypsin treatment *in vitro* into the biologically active F protein consisting of the F_1 and F_2 subunits, while cleavage occurs naturally in eggs (Homma and Ouchi 1973; Scheid and Choppin 1974). Therefore, SeV cannot undergo multiple cycles of infection or produce plaques in cell culture unless trypsin is added to the culture medium. Its tissue tropism *in vivo* depends upon whether an activating protease is available in the tissue. This discovery of cleavage activation of SeV was a clue to establishing the general concepts shared with many enveloped viruses in different virus families that the envelope protein responsible for initiation of infection through fusion of the viral envelope with the target cell membrane is synthesized as a biologically inactive precursor and converted to an active form by proteolytic cleavage and that the availability of activating protease is a key determinant of viral tropism and pathogenicity for many of those viruses. This concept of protease-dependent viral tropism originated from early studies, which compared virulent NDV strains sensitive to a ubiquitous processing protease and causing a pantropic (systemic) lethal infection and avirulent NDV strains insensitive to a ubiquitous protease but cleavable by a highly tissue-specific protease and causing a localized benign infection (Nagai et al. 1976b; Toyoda et al. 1987, 1989; Ohnishi et al. 1995) and extended to other enveloped viruses including influenza A virus (Nagai 1993, 1995; Klenk and Garten 1994).

As there is a single arginine residue at the cleavage site of SeV F_0, it is not cleavable by a ubiquitous intracellular, *trans*-Golgi localized endoprotease such as furin that preferentially recognizes and cleaves a multibasic or paired basic motif (Nagai 1993, 1995). Accordingly, SeV F_0 is cleaved by an extracellular protease. As a naturally occurring SeV-activating endoprotease, the blood clotting factor Xa (FXa) was first isolated from the allantoic and amniotic fluids of chicken eggs (Gotoh et al. 1990; Suzuki et al. 1991; Ogasawara et al. 1992). A protease with similar substrate specificity to that of FXa, tryptase Clara, was then suggested to be a responsible protease for SeV activation on the epithelial surface of the respiratory tract of rats (Kido et al. 1992; Tashiro et al. 1992). More recently, it was suggested that mini-plasmin expressed in the bronchioles and anionic trypsin ectopically expressed in the alveoli of rats (Kido et al. 2007), as well as the serine transmembrane proteases,TMPRSS2, TMPRSS4, and HAT, on human airways (Böttcher et al. 2006; Shirogane et al. 2008; Chaipan et al. 2009), fall in the same category of virus-activating protease.

2.5.7 *Assembly, Budding, and Release*

2.5.7.1 Intracellular Trafficking of the RNPs to the Site of Assembly

Molecular imaging technology is now applied widely from the cellular level to the entire organism to study trafficking and localization of the molecules of interest. To analyze real-time trafficking of SeV RNP in a living cell, Chambers and Takimoto (2010) created an rSeVLeGFP, which expresses the L protein fused to enhanced green fluorescent protein (eGFP) and showed similar growth kinetics to that of the wild-type SeV. The majority of LeGFP colocalized with the RNPs and N and P proteins, suggesting that the fluorescent signals could represent the locations of RNPs. Analysis of LeGFP movement with a time-lapse digital video microscopy revealed directional and salutatory movement of LeGFP along microtubules. This movement was restricted by nocodazole (an inhibitor of microtubule polymerization) treatment, which resulted in reduced production of progeny virions, suggesting the role of microtubules in RNP trafficking and virus assembly. With electron microscopy, the RNPs were found to be closely associated with the intracellular vesicles. Further, colocalization of RNPs and Rab11a protein, which is an RAS oncogene-related small GTPase protein and known to regulate the recycling of the endocytosis pathway and Golgi-to-plasma membrane trafficking, and simultaneous movement of LeGFP and Rab11a were observed, suggesting the involvement of vesicle recycling in RNP translocation. These data suggested that SeV RNPs traffic through the cytoplasm to the site of assembly using the intracellular vesicles along the microtubules network. Earlier *in vitro* studies using purified virions of SeV and VSV suggested that microtubules were necessary for the viral RNA synthesis (Moyer et al. 1986). Thus, the microtubules appear to be involved in both SeV RNA synthesis and RNP transportation in cells.

2.5.7.2 The Roles of M Protein in Assembling Viral Components and Driving Budding

All viral structural components of a paramyxovirus meet at the plasma membrane and are organized into a mature virion, which buds from the plasma membrane (Fig. 2.2). For SeV and many paramyxoviruses the M protein has been thought to be the central player in this assembly and budding process. How the M protein plays this role has been a central question. Earlier studies by *in vitro* reconstitution of isolated SeV glycoproteins, M protein and RNP demonstrated that association of the glycoproteins with the RNP occurred only if M protein was present (Shimizu and Ishida 1975; Yoshida et al. 1976), suggesting that bridging the viral envelope and the viral core is one of the roles played by the M protein. The SeV M protein was demonstrated to interact with the cytoplasmic tails of F and HN proteins in the absence of the other viral components including RNP (Ali and Nayak 2000). These interactions were SeV specific because no interaction occurred when influenza virus glycoproteins were used in place of the homologous SeV glycoproteins. Using freshly prepared and thus structurally firm virions (see Sect. 2.3) and chemical crosslinkers, Markwell and Fox (1980) demonstrated that the M proteins lie in close proximity (within 1.1 nm) to the RNP. Thus, virus-specific physical HN/F–M–RNP interactions appear to have been well established (Nagai et al. 2011).

Straightforward evidence indicating the critical requirement of M protein for SeV assembly and budding came from studies showing that the virus particle formation was completely abrogated in cells infected with an rSeV with the *M* gene deleted (Δ*M*SeV), whereas *HN* or *F* gene deletion permitted particle formation (Inoue et al. 2003b; see Chap. 3). This observation was complemented by the results that suppression of the M-protein expression by an siRNA during the viral lifecycle also abrogated virion formation (Mottet-Osman et al. 2007).

The SeV M protein contains five cysteine residues at positions 83, 106, 158, 251, and 295, all of which are not involved in disulfide bond formation but are present in a free form throughout the lifecycle (see Sect. 2.5.6.2). Three rSeVs (Cys83Ser,

Cys106Ser, and Cys295Ser) were successfully generated, and the virus particles collected from culture media of infected cells were examined for their size distributions (Sakaguchi et al. 2002). The parental virus displayed a typical normal distribution pattern with a sharp peak around 175–224 nm in diameter; 68.27 % of virus particles were within 200 ± 48 nm in diameter (see Sect. 2.3.1). In contrast, the patterns of the three mutants largely deviated from the normal distribution. Cys83Ser and Cys106Ser yielded smaller-size virions of varying diameters, 173 ± 57 and 157 ± 46 nm, respectively, whereas Cys295Ser was larger and highly size heterogeneous (213 ± 96 nm). Moreover, Cys106Ser yielded a significant amount of empty particles lacking the RNPs. These results further substantiate the critical role of M protein in SeV morphogenesis.

At least a portion of intracellular M proteins was found to be phosphorylated although those in the virions were not (Lamb and Choppin 1977b), which suggested the possibility that the phosphorylated M was selectively eliminated from progeny virions during the process of assembly or that M was dephosphorylated during the process. After identifying the serine residue at residue 70 as the phosphorylation site, an rSeV Ser70Ala was generated that no longer underwent M protein phosphorylation. This mutant exhibited no particular phenotype in cell cultures (growth kinetics and virus yield) and in mice (killing ability). Thus, it appeared that phosphorylation of SeV M protein is just an epiphenomenon unrelated to either the lifecycle or pathogenesis (Sakaguchi et al. 1997).

In the final step of the lifecycle, the glycoproteins of SeV and NDV appear to be initially dispersed over the entire cell surface and then condensed into a patch of viral membrane with morphologically visible spikes in the regions of membrane where the M proteins are present (Nagai et al. 1976a, 1983b). The RNPs align beneath those modified membrane areas. Also, the membrane sites to which RNPs adhere are modified by virus-specific particles arranged in a crystalline orthogonal array (Bächi 1980), which strikingly resembles the cylindrical structure formed by the M proteins (Hewitt and Nermut 1977; Heggeness et al. 1982). However, it is not known whether the association of M protein promotes condensation of the glycoproteins into a patch or if patch formation is a precondition for attachment of M protein. The former possibility appeared to be supported by the studies with an SeV mutant, which was temperature sensitive in M-protein synthesis (Yoshida et al. 1979). The glycoproteins exhibited much higher mobility in the plane of the membrane, as detected by glycoprotein capping by the specific antibody, in the absence of M protein synthesis at a nonpermissive temperature than when M-protein synthesis was allowed to occur at a permissive temperature.

Similar to other enveloped viruses, SeV proteins (F, HN, M, N, and P) became, at least in part, associated with a mild detergent-resistant fraction, the lipid raft, which is thought to be a platform of virus assembly (Gosselin-Grenet et al. 2006). In one study, it was found that when expressed alone by transfection, the HN, F, or M protein independently became partially detergent resistant (Sanderson et al. 1995), while in another similar transfection experiment, M protein was found associated with the detergent-resistant fraction only when coexpressed with the HN and F proteins (Ali and Nayak 2000).

Thus, although the critical requirement of the M protein for SeV assembly and budding appears to have been well established, there remain many issues to be settled. It needs to be determined, for example, whether patch formation occurs in Δ*M*SeV-infected cells and if not, whether *trans*-supplied M proteins promote patch formation. Δ*M*SeV is also useful for determining whether the glycoproteins would become associated with the lipid raft in the absence of M protein (Nagai et al. 2011).

2.5.7.3 The C Proteins as an Additional Player

The SeV accessory C proteins are essentially nonstructural proteins because they are detectable in very few copies in the virions. It is thus surprising that the C proteins likely contribute greatly to the structural assembly process: this was first suggested by the experiments using 4C(−) virus that expressed none of the 4C proteins (see Sect. 2.5.5.3). When propagated in eggs, 4C(−) replicated to the titers several logs lower than that of the wild type (Kurotani et al. 1998). In CV1 cell culture, the peak titer of 4C(−) was about 2 logs lower than that of the wild type, despite efficient synthesis of mRNAs, proteins, and genomic RNAs (Hasan et al. 2000), although these features appeared to result, at least in part, from the genome polarity shift toward the predominance of (+) RNA in cells and virions (see Sect. 2.5.3.3). In addition, the 4C(−) virions produced were highly anomalous in size and shape (Hasan et al. 2000).

The VLP (virus-like particle) production in cells transfected with the plasmids encoding the N, M, HN, and F proteins or with the M protein-encoding plasmid alone was enhanced by an additional expression of the C protein (Sugahara et al. 2004; Irie et al. 2008b). The longer C proteins (C′ and C) contain N-terminal membrane-targeting sequences, whereas the shorter Y1 and Y2 lack these sequences (Marq et al. 2007), and the former two, but not the latter two, were capable of facilitating VLP production (Sakaguchi et al. 2005; Irie et al. 2008b).

However, such a supporting role of C protein in SeV budding was not confirmed by others (Gosselin-Grenet et al. 2007). The rSeV used in this experiment was a 4C knockout virus into which a separate transcription unit encoding the C protein fused with GFP was introduced. Even when the C–GFP expression was silenced by siRNA, no apparent inhibition of the production of the rSeV was found. However, siRNA-induced depletion might not be sufficient for inhibiting virus production, as about 30 % of the C–GFP protein level, compared with that of the control cells, remained expressed (Nagai et al. 2011).

2.5.7.4 Conflicting Results Regarding the Involvement of ESCRT Machinery

Evidence is accumulating that suggests that the matrix proteins of many enveloped viruses such as the paramyxovirus M protein and retrovirus Gag protein utilize the cellular endosomal sorting complexes required for transport (ESCRTs), which are involved in multi-vesicular body formation and cytokinesis, for the final (late) step of virus assembly (release of the budding particles from the cellular membrane by the process of fission). Three major viral L (late)-domain motifs, ProProXaaTyr (PPxY), ProThr/SerAlaPro (PT/SAP), and TyrProXaaLeu (YPxL), have been identified as the responsible sequences for that interaction with the ESCRTs (Demirov and Freed 2004; Bieniasz 2006; Calistri et al. 2009; Harrison et al. 2010).

A potential L-domain 49-TyrLeuAspLeu (YLDL)-52 was identified for SeV M protein, and its interaction with Alix/AIP1 (at the N-terminal residues 1–211) appeared to be important for the formation of VLPs from plasmid-based expression of SeV M proteins (Irie et al. 2007). The Alix is associated with the two (ESCRT-I and -III) of the three ESCRTs and was demonstrated to be the interacting partner of the lentivirus YPxL (Demirov and Freed 2004). Although SeV YLDL could not be replaced by YPxL, SeV M-protein budding was inhibited by the overexpression of some deletion mutants of Alix and depletion of endogenous Alix by siRNA (Irie et al. 2007), whereas overexpression of Alix enhanced SeV virion release from infected cells (Sakaguchi et al. 2005). An rSeV with point mutations in the YLDL motif replicated poorly in cells, and a revertant that emerged restored replication capacity (Irie et al. 2010).

The SeV C protein, which was suggested to play a role in virus assembly using 4C(−) virus (see Sect. 2.5.7.3), was also found to physically interact with Alix (at the C-terminal residues 212–357) (Sakaguchi et al. 2005; Irie et al. 2007). The mutant C proteins that were unable to bind Alix (Sakaguchi et al. 2005) or that failed to bind the plasma membrane (Irie et al. 2008b) could not accelerate the release of VLP from M-, N-, HN-, and F-expressing cells or from only M protein-expressing cells. Thus, SeV M and C proteins appeared to synergistically facilitate M-protein budding through the interaction with different Alix domains.

In the final step of protein sorting, AAA-type ATPase, Vps4 interacts with ESCRT-III to catalyze disassembly of the ESCRT machinery to recycle its components (Babst et al. 1998). Dominant-negative (DN) Vps4 mutants interfered with the budding of many enveloped viruses including parainfluenza virus 5 (PIV5) and mumps virus in the genus *Rubulavirus* (Schmitt et al. 2005; Li et al. 2009; reviewed in Bieniasz 2006 and Harrison et al. 2010). Sakaguchi et al. (2005) reported that transfection of a DN Vps4 yielded significantly lower amounts of infectious SeV in cells transfected with infectious SeV nucleocapsid. DN Vps4 protein expression was found to affect the release of SeV M protein VLP that occurred in the presence of the C-protein expression, but not M-alone VLP production (Irie et al. 2008b).

Overall, the foregoing results suggested that SeV budding and release from cells involve recruitment of Alix and ESCRT machinery by the viral M and C proteins. However, in contrast to those "positive" results, Gosselin-Grenet et al. (2007) presented "negative" data that in the SeV natural lifecycle neither the depletion of Alix expression nor DN Vps4 expression affected virus production. In this study on DN Vps4 expression, it has to be noted that the experimental system appeared to be well validated in a parallel setting of PIV5 whose replication was demonstrated to depend upon Vps4 and inhibited by a DN Vps4 (Schmitt et al. 2005). Thus, currently available concepts are sharply conflicting regarding the involvement of ESCRT machinery in SeV assembly and budding (Harrison et al. 2010; Nagai et al. 2011).

As already described, a VLP setup has very often been used for studies on assembly and budding. However, it always has to be taken into consideration that the setup does not always faithfully reflect the natural assembly and budding process. For example, tetherin, which is expressed at the plasma membrane and is localized to lipid rafts, restricted the release of Ebola virus VLPs but not Ebola virus itself (Radoshitzky et al. 2010). Similarly, influenza virus growth was not restricted by tetherin but influenza VLP production was restricted (Watanabe et al. 2011). Thus, what has been revealed by the VLP setup needs to be verified in the context of the natural lifecycle.

2.5.7.5 Release

After fission of the bud, the virus is liberated from cells and initiates the next round of the lifecycle in a new cell (Fig. 2.2). The NA of the HN glycoprotein removes the terminal sialic acid residues from the HN glycoprotein itself and the F glycoprotein, thereby preventing particle aggregation and facilitating virus release from cells. The NA has acidic pH optima and perhaps acts primarily in the Golgi apparatus. However, the erythrocytes once agglutinated by the exogenous SeV readily become liberated in hemagglutination tests at neutral pH, even on ice, which suggests that the SeV NA would further act exogenously, for instance, to remove cell-surface sialic acids to prevent reattachment of progeny virus.

2.6 SeV-Persistent Infection: Two-Step Theory for Its Establishment

A number of paramyxoviruses can establish long-term persistent infections in cultured cells (Rima and Martin 1976; Youngner and Preble 1980). Paramyxovirus-persistent infections are also established *in vivo*, and the resulting host responses and pathological outcomes are complex (Randall and Russell 1991). A typical example of human disease caused by a persistently infecting paramyxovirus is subacute sclerosing panencephalitis (SSPE), a disease of the human central nervous system, which is triggered by persistently infecting measles virus many years after the onset of acute measles. Despite its very rare occurrence, SSPE has been intensively studied to reveal the molecular bases underlying the pathogenesis (Billeter and Cattaneo 1991).

Although its clinical impact is not known, SeV has been intensively used to understand the process leading to the establishment of a stable persistent infection at the cellular level. Roux and Holland (1979, 1980) showed that SeV-persistent infection could be established by coinfection with a standard, nondefective (ND) virus and DI particles. DI RNAs persisted in the cells and maintained their interfering capacity with the replication of ND virus. On the other hand, temperature-sensitive (ts) ND mutants were frequently isolated from cells persistently infected with SeV (Nagata et al. 1972). The ts mutants grew well at a permissive temperature (33 °C), but did not grow at a nonpermissive temperature (38 °C), although the majority of cells expressed viral antigens at the nonpermissive temperature. Such SeV ts mutants were capable of establishing a persistent infection (Kimura et al. 1975). Thus, two factors, presence of DI particles and evolution of ts mutants, have been thought to be critical for establishing a stable SeV-persistent infection (Nagai et al. 2011).

SeV causes lytic infection for cultured cells. Routinely, a small fraction of cells that survived this lethal infection have to be carefully cultivated and expanded to initiate persistent infection. They are then subjected to numerous rounds of passage until they become stably maintained (steady state) and completely resistant to the original SeV. During serial passages, however, these cells encounter occasional crisis (massive cell death). Yoshida et al. (1982) characterized SeV genomes that persisted in a stable culture established after numerous rounds of passages and found that both full-length and smaller-sized DI genomes coexisted. In addition, a temperature-sensitive (ts) mutant isolated from such stable lines was found to be able to immediately establish a stable persistent infection without detectable DI genomes. Thereafter, no detectable level of DI genome evolved and the cells no longer encountered sudden crisis. Because of the widespread occurrence of DI particles among RNA viruses, the stock SeV initially used could contain a trace level of DI particles; this might be sufficient for protecting a small cell population from lytic infection and thus initiating a persistent infection. DI particles require standard wild-type virus to reproduce themselves. Thus, the persistent infection initiated by the aid of DI particles could be maintained by a subtle balance of intracellular levels of lytic wild-type and non-lytic DI particles. Once such a balance shifts toward the predominance of wild type, the system probably undergoes crisis. Persistent infection then enters the steady state no longer showing crisis. Just concurrently with, but never antecedently to, this stable phase, ts mutants began to arise. They were always slightly cytopathic, multiplied without wild type, and, importantly, interfered strongly with the replication of the wild type (Yoshida et al. 1982; Kiyotani et al. 1990). Once evolved, the ts mutants could replace the wild type and ultimately expel it from the system. Thus, a two-step theory has emerged that SeV-persistent infection is initiated by DI particles followed by stabilization by endogenously generated noncytopathic ts mutants with homologous interfering capacity (Nagai and Yoshida 1984; Nagai et al. 2011).

The nucleotide sequence of a persistent infection-derived ts mutant (named cl. 151) with homologous interfering capacity and noncytopathic nature was compared with that of the parental SeV strain Nagoya. Numerous nonsynonymous and synonymous mutations were identified throughout the cl. 151 genome (Nishimura et al. 2007). The two nucleotide changes, A to G at position 34 and G to A at position 47, were found in the leader sequence. These two changes were introduced into the parental virus. The resulting rSeV displayed increased homologous-interfering capacity (Shimazu et al. 2008). The Leu1618Val mutation in the L protein of the ts mutant also appeared to be relevant to the establishment of persistent infection but did not confer temperature sensitivity (Nishio et al. 2004). On the other hand, several amino acid changes in the M and HN proteins of cl. 151 (Kondo et al. 1993) were found to contribute to its temperature sensitivity as well as poor cytopathogenicity (Inoue et al. 2003a).

2.7 Indispensability for *In Vivo* Pathogenesis of the Accessory V and C Proteins

SeV causes fatal pneumonia in mice. When intranasally inoculated under our experimental conditions, the virus titers in the lung increase exponentially during the initial 1 or 2 days; this is followed by a period of high viral load of up to 7–9 days, during which the lung lesion (consolidation) score increases (see the wild type in the $STAT1^{+/+}$ control mice in Fig. 2.4). All mice eventually die of severe hemorrhagic pneumonia. On the other hand, V-knockout [V(−) and VΔC] and C-knockout [4C(−)] viruses (see Sect. 2.5.5.3) were unable to maintain a high viral load because they could not grow at all in the mouse lung [4C(−)] (Kurotani et al. 1998) or did grow but were rapidly cleared from the lung [V(−) and VΔC] (Kato et al. 1997a, b) and therefore displayed no fatality. Thus, V and C proteins are critically important for counteracting host effectors to limit the virus spreading in the body. Thus, the paramyxovirus accessory genes have become a focus of active investigation by reverse genetics (Nagai and Kato 2004). A great deal of effort has been made to identify those host effectors targeted by the SeV V and C accessory proteins at the

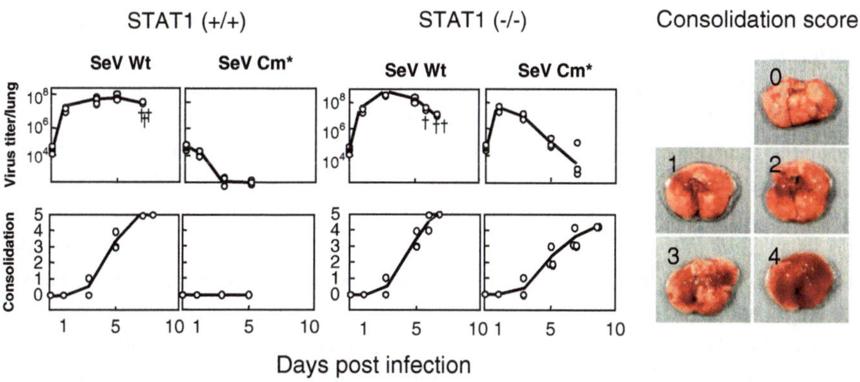

Fig. 2.4 Growth and pathogenicity of SeV Cm* in wild-type and $STAT1^{-/-}$ mice. Five-week-old wild-type $STAT1^{+/+}$ 129S6 and STAT1-deficient ($STAT1^{-/-}$) 129S6 mice were infected intranasally with 10^7 CIU (cell infection units) SeV Wt or SeV Cm*. Two or three mice from each group were killed at the intervals indicated and examined for virus infectivity in the lung (*top*) and lung consolidation (*bottom*). *Dagger* symbol indicates a dead mouse. Actual macroscopic views of the lung corresponding to the consolidation scores of 0 (no lesion) through 4 (maximum) are presented as references (*right*). When a mouse died, one point was added to the consolidation score (score 5). (Adapted from Sakaguchi et al. 2008)

cellular level, revealing a variety of potential targets (reviewed in Sakaguchi et al. 2008; Nagai et al. 2011). Here, we describe only those targets whose significance has been verified *in vivo* using a variety of mutant mice.

2.7.1 Blocking the IFN-α/β Signaling by the C Protein

Infection with SeV and other paramyxoviruses has been known to render cells unresponsive to IFNs (IFN-α/β and IFN-γ), allowing highly IFN-sensitive viruses such as VSV to replicate in the presence of IFNs.

Host cells sense virus invasion by recognizing microbe-specific pathogen-associated molecular patterns (PAMPs), such as double-stranded RNA generated in the course of infection and, in particular for NNSVs, the 5′-triphosphate end of genomic RNA generated during replication (Hornung et al. 2006), and such recognitions activate the signaling pathway leading to IFN-β production followed by the amplification of IFN-α and IFN-β (IFN-α/β). Autocrine or paracrine IFN-α/β binds to the cell-surface receptors, and its signal is transmitted to the nucleus through the IFN signaling (JAK/STAT) pathway, resulting in activation of IFN-stimulated genes (ISGs). The products of ISGs include antiviral proteins, such as dsRNA-dependent protein kinase (PKR), which participate in establishment of the antiviral state.

In SeV-infected cells, neither IFN-α/β nor IFN-γ can activate transcription of various ISGs and a reporter gene under the control of the IFN-responsive promoter (Didcock et al. 1999a; Garcin et al. 1999; Gotoh et al. 2001, 2002), indicating SeV ability to prevent IFN signaling. Using V-knockout and C-knockout rSeVs, studies by Garcin et al. (Garcin et al. 1999) and Gotoh et al. (Gotoh et al. 1999) sought to identify the viral factors responsible for the blockade of IFN signaling. Of the three different rSeVs, V(−), C/C′(−), and 4C(−), only 4C(−) completely lost the ability to prevent IFN signaling and permitted the induction of the full antiviral state (Gotoh et al. 1999). Plasmid-based expression of the individual proteins indicated that not only the longer C and shorter Y1 and Y2, but also the C-terminal half of the C protein (aa 99–204), was able to inhibit IFN-mediated induction of ISGs and subsequent establishment of the antiviral state, whereas expression of V protein did not do so at all (Kato et al. 2001, 2002, 2004; Gotoh et al. 2003a, b). It was concluded that the C-terminal half of the SeV C protein encodes the IFN antagonism (reviewed in Nagai and Kato 2004).

The *Rubulavirus* V proteins target STAT1 or STAT2 for proteasome-mediated degradation and thereby block IFN signaling (Didcock et al. 1999b; Randall and Goodbourn 2008). In contrast, no degradation of signaling molecules including STAT1 and STAT2 was observed for not only SeV-infected HeLa cells, but also the established HeLa cell lines constitutively expressing the C, Y1, or Y2 (Komatsu et al. 2000; Young et al. 2000; Kato et al. 2001). Instead, phosphorylation of STAT1 and STAT2 was markedly affected (Komatsu et al. 2000, 2002; Gotoh et al. 2003b). After continuous stimulation with IFN-α in the presence of SeV C protein in cells, tyrosine phosphorylation of STAT1 was initiated first with slower kinetics, but eventually reached a high level (Komatsu et al. 2000, 2002), while tyrosine phosphorylation of STAT2 remained almost completely suppressed (Gotoh et al. 2003b). Analysis of established HeLa cell lines, each of which constitutively expresses a truncated C or a mutated C with substitutions of one to three amino acids, revealed a strict correlation between the ability to block IFN-α/β signaling and the ability to inhibit STAT2 phosphorylation (Gotoh et al. 2003b; Kato et al. 2004), demonstrating that inhibition of tyrosine phosphorylation of STAT2 was a crucial step for SeV-mediated blockade of IFN-α/β signaling (reviewed in Sakaguchi et al. 2008; Nagai et al. 2011).

By introducing three amino-acid substitutions into the C protein, an rSeV (Cm*) was generated. Cm* was found to have lost the ability to prevent IFN-α/β signaling, but still retained the other functions including RNA synthesis inhibition and facilitation of virus budding (Kato et al. 2007).

When intranasally inoculated, Cm* could propagate in the lungs of $STAT1^{-/-}$ mice and cause pneumonia, but was cleared from the lungs of $STAT1^{+/+}$ mice without any appreciable propagation and pathogenicity (Fig. 2.4) (Kato et al. 2007). Thus, the critical requirement of virus-coded IFN antagonism for *in vivo* replication and pathogenesis was proved for the first time with the SeV C protein (reviewed in Sakaguchi et al. 2008; Nagai et al. 2011).

The SeV C proteins were further found to suppress IFN-β production, antagonize IFN-γ signaling, and encode anti-apoptotic function (reviewed in Nagai et al. 2011). The significance of these findings made *in vitro* remains to be verified *in vivo*.

2.7.2 Blocking by the V Protein of Innate Virus Clearance Through IRF3 Activation but via a Yet Unknown Pathway

Both V(–) and VΔC proliferated in the lungs of normal mice as efficiently as the parental wild-type SeV until day 1 post infection. However, they were then rapidly cleared and produced only moderate consolidation scores without killing the mice, in contrast to the wild-type SeV, which maintained both a high viral load and high consolidation scores up to 7–9 days and killed all the mice (Kato et al. 1997a, b) (also see $IRF3^{+/+}$ mice in Fig. 2.5). A similar attenuation was found for several rSeVs with the zinc finger-like motif in the V-unique region disrupted and without the capacity to bind zinc (Huang et al. 2000; Fukuhara et al. 2002). The V protein, which is fully dispensable in the viral lifecycle, was therefore found to encode a luxury function required for pathogenesis. The V-unique region, particularly its zinc-binding capacity, appeared to be crucial for this function. Exactly the same phenotypic differences were found between the V(–) and wild type in NK cell-deficient beige mice, T cell-deficient nu/nu mice, and the mouse strains deficient in NKT, TNF-α, IL6, *i*NOs, TRIF, and MyD88 (Kiyotani et al. 2007, manuscript in preparation; reviewed in Sakaguchi et al. 2008; Nagai et al. 2011).

Remarkably, however, $IRF3^{-/-}$ mice were found to be highly and almost equally susceptible to V(–), VΔC, and wild-type viruses (Kiyotani et al. 2007). The unique,

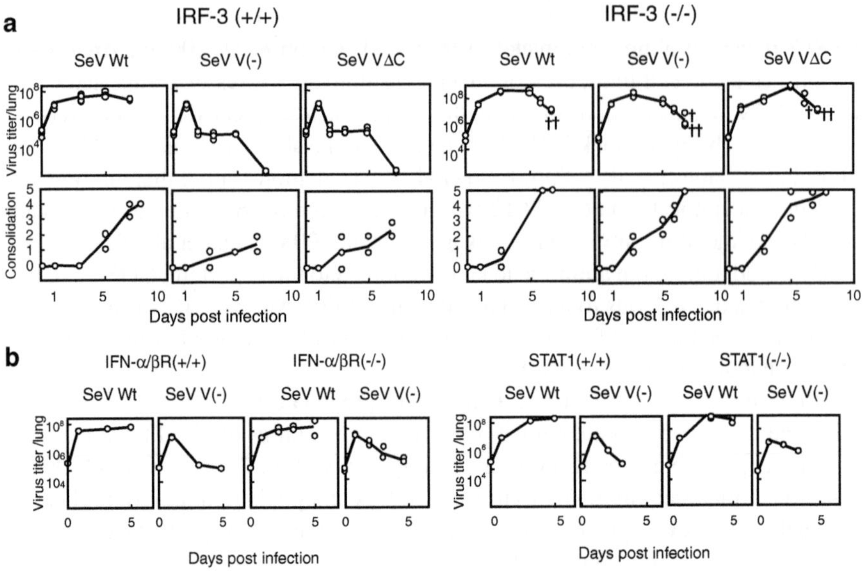

Fig. 2.5 Replication and pathogenicity of SeV V(−) and SeV VΔC in IRF3$^{-/-}$ mice. (**a**) Five-week-old IRF3$^{+/+}$ C57BL/6J and IRF3-deficient (IRF3$^{-/-}$) C57BL/6J mice were infected intranasally with 10^7 CIU SeV Wt, SeV V(−), or SeV VΔC. Two or three mice from each group were killed at the indicated intervals and examined for virus infectivity in the lung and lung consolidation. *Dagger* symbol indicates a dead mouse. (**b**) Six- to 7-week-old IFN-α/β R$^{+/+}$ A129/Sv and IFN-α/β receptor-deficient (IFN-α/β R$^{-/-}$) A129/Sv mice were infected intranasally with 10^7 CIU SeV Wt or SeV V(−) (*left*). Five-week-old STAT1$^{+/+}$ 129S6 and STAT1-deficient (STAT1$^{-/-}$) 129S6 mice were infected intranasally with 10^7 CIU SeV Wt or SeV V(−) (*right*). Viral infectivity in the lung was measured at the indicated time intervals after infection. (Adapted from Sakaguchi et al. 2008)

rapid clearance of V(−) and VΔC viruses could no longer be observed. Instead, they grew as efficiently as the wild type and caused severe pneumonia and death as rapidly as the wild type (Fig. 2.5). The SeV V protein was recently found to inhibit IRF3 activation and its nuclear translocation (Irie et al. 2012). IFN-β is one of the well-known IRF3-inducible gene products, and IFN-α/β is subsequently amplified via IRF7 activation. In agreement with this, IFN-α/β production was increased by several fold in normal mice infected with V(−) or VΔC, compared with wild-type infection. It was therefore a reasonable assumption that the SeV V protein would also be a type I IFN (IFN-α/β) antagonist. However, no such active growth beyond 1 day after infection as seen in IRF3$^{-/-}$ mice was observed for the V(−) virus in IFN-α/β receptor$^{-/-}$ mice and STAT1$^{-/-}$ mice, where the IFN-α/β-mediated antiviral state is not inducible (Kiyotani et al. 2007) (Fig. 2.5). V(−) appeared to be cleared in these mice as efficiently as in normal mice. The V(−) virus was also efficiently cleared in IFN-γ$^{-/-}$ mice, indicating that type II IFN is also not involved (Kiyotani et al. 2007; reviewed in Sakaguchi et al. 2008; Nagai et al. 2011).

What IRF3-induced effectors other than the IFN system are responsible for the rapid clearance of V(−) and VΔC remains to be elucidated. In this context, it is noteworthy that activated IRF3 up-regulates some chemokine genes such as RANTES, and other ISGs, and further that the IRF3/Bax pathway is involved in SeV-induced apoptosis (Heylbroeck et al. 2000; Peters et al. 2008; Chattopadhyay et al. 2010).

2.8 Conclusions

SeV has been an irreplaceable tool in basic research to understand paramyxovirus replication and pathogenesis since its discovery six decades ago. Many of the paramyxovirus theories and traits revealed find their origin in SeV research. In fact, Brian W.J. Mahy at the Centers for Disease Control and Prevention (GA, USA) stated in the introduction of a recent textbook (*The Biology of Paramyxoviruses*, ed. S.K. Samal, Caister Academic Press, Norwalk, UK, 2011) that a great deal of basic research to understand paramyxovirus replication and pathogenesis has been carried out using SeV and that SeV has contributed more to our understanding of paramyxoviruses than any other members of the family. However, the research outcomes are still at the phenomenological level and sometimes even in serious conflict, as illustrated in some sections. Thus, continued efforts are needed for better understanding of SeV and paramyxovirus biology.

For certain paramyxoviruses and other NNSV species, reverse genetics has evolved into technology to use these viruses as an expression vector. Studies to develop the SeV vector appear to typify this line of research because they have demonstrated great potential of diverse applications and because some of them have now left or are about to leave the laboratory bench for preliminary clinical trials (see subsequent chapters). Active incorporation of basic knowledge on SeV biology described here has accelerated the production of diverse target-oriented devices to be used in special settings (also see subsequent chapters). Such progress in SeV technology appears to be largely the result of some biological and ecological characteristics intrinsic to this creature that were highlighted here and in the subsequent chapters, including nonintegrating nature, extremely high production titers, extremely high performance of transgenes expression, remarkable target cell breadth, and no potential pathogenicity for humans. The clinical as well as preclinical research outcomes will be returned to basic biology. It is our hope that SeV basics and technology spiral up, closely interacting with each other.

References

Ali A, Nayak DP (2000) Assembly of Sendai virus: M protein interacts with F and HN proteins and with the cytoplasmic tail and transmembrane domain of F protein. Virology 276:289–303

Babst M, Wendland B, Estepa EJ, Emr SD (1998) The Vps4p AAA ATPase regulates membrane association of a Vps protein complex required for normal endosome function. EMBO J 17:2982–2993

Bächi T (1980) Intramembrane structural differentiation in Sendai virus maturation. Virology 106:41–49

Bieniasz PD (2006) Late budding domains and host proteins in enveloped virus release. Virology 344:55–63

Billeter MA, Cattaneo R (1991) Molecular biology of defective measles viruses persisting in the human central nervous system. In: Kingsbury DW (ed) The paramyxoviruses. Plenum Press, New York

Bitzer M, Lauer U, Baumann C, Spiegel M, Gregor M, Neubert WJ (1997) Sendai virus efficiently infects cells via the asialoglycoprotein receptor and requires the presence of cleaved F0 precursor proteins for this alternative route of cell entry. J Virol 71:5481–5486

Blanchard L, Tarbouriech N, Blackledge M, Timmins P, Burmeister WP, Ruigrok RWH, Marion D (2004) Structure and dynamics of the nucleocapsid-binding domain of the Sendai virus phosphoprotein in solution. Virology 319:201–211

Bossow S, Schiecht S, Schubbert R, Pfeiffer M, Neubert WJ, Wiegand M (2012) Evaluation of nucleocapsid and phosphoprotein P functionality as critical factors during the early phase of paramyxoviral infection. Open Virol J 6:73–81. doi:10.2174/1874357901206010073

Böttcher E, Freuer C, Steinmetzer T, Klenk HD, Garten W (2006) MDCK cells that express proteases TMPRSS2 and HAT provide a cell system to propagate influenza viruses in the absence of trypsin and to study cleavage of HA and its inhibition. Vaccine 27:6324–6329

Bousse T, Takimoto T (2006) Mutation at residue 523 creates a second receptor binding site on human parainfluenza virus type 1 hemagglutinin-neuraminidase protein. J Virol 80: 9009–9016

Buchholz CJ, Spehner D, Drillien R, Neubert WJ, Homann HE (1993) The conserved N-terminal region of Sendai virus nucleocapsid protein NP is required for nucleocapsid assembly. J Virol 67:5803–5812

Cadd T, Garcin D, Tapparel C, Itoh M, Homma M, Roux L, Curran J, Kolakofsky D (1996) The Sendai paramyxovirus accessory C proteins inhibit viral genome amplification in a promoter-specific fashion. J Virol 70:5067–5074

Calain P, Roux L (1993) The rule of six, a basic feature for efficient replication of Sendai virus defective interfering RNA. J Virol 67:4822–4830

Calistri A, Salata C, Parolin C, Palù G (2009) Role of multivesicular bodies and their components in the egress of enveloped RNA viruses. Rev Med Virol 19:21–45

Çevik B, Kaesberg J, Smallwood S, Feller JA, Moyer SA (2004) Mapping the phosphoprotein binding site on Sendai virus NP protein assembled into nucleocapsids. Virology 325:216–224

Çevik B, Smallwood S, Moyer SA (2007) Two N-terminal regions of the Sendai virus L RNA polymerase protein participate in oligomerization. Virology 363:189–197

Chaipan C, Kabasa D, Bertram S, Glowacka I, Steffen I, Tsegaye TS, Takeda M, Bugge TH, Kim S, Park Y, Marzi A, Pöhlmann S (2009) Proteolytic activation of the 1918 influenza virus hemagglutinin. J Virol 83:3200–3211

Chambers R, Takimoto T (2010) Trafficking of Sendai virus nucleocapsids is mediated by intracellular vesicles. PLoS One 5:e10994. doi:10.137/journal.pone.0010994

Chandrika R, Horikami SM, Smallwood S, Moyer SA (1995) Mutations in conserved domain I of the Sendai virus L polymerase protein uncouple transcription and replication. Virology 213:352–363

Chattopadhyay S, Marques JT, Yamashita M, Peters KL, Smith K, Desai A, Williams BR, Sen GC (2010) Viral apoptosis is induced by IRF-3-mediated activation of bax. EMBO J 29:1762–1773

Choppin PW, Compans RW (1975) Reproduction of paramyxoviruses. In: Fraenkel-Conrat H, Wagner RR (eds) Comprehensive virology, vol 4. Plenum Press, New York

Cortese CK, Feller JA, Moyer SA (2000) Mutations in domain V and the Sendai virus L polymerase protein uncouple transcription and replication and differentially affect replication in vitro and in vivo. Virology 277:387–396

Curran J (1996) Reexamination of the Sendai virus P protein domains required for RNA synthesis: a possible supplemental role for the P protein. Virology 221:130–140

Curran J (1998) A role for the Sendai virus P protein trimer in RNA synthesis. J Virol 72:4274–4280

Curran J, Kolakofsky D (1987) Identification of an additional Sendai virus non-structural protein encoded by the P/C mRNA. J Gen Virol 68:2515–2519
Curran J, Kolakofsky D (1988a) Ribosomal initiation from an ACG codon in the Sendai virus P/C mRNA. EMBO J 7:245–251
Curran J, Kolakofsky D (1988b) Scanning independent ribosomal initiation of the Sendai virus X protein. EMBO J 7:2869–2874
Curran J, Kolakofsky D (1989) Scanning independent ribosomal initiation of the Sendai virus Y proteins in vitro and in vivo. EMBO J 8:521–526
Curran J, Kolakofsky D (2008) Nonsegmented negative-strand RNA virus RNA synthesis in vivo. Virology 371:227–230
Curran J, Boeck R, Kolakofsky D (1991) The Sendai virus P gene expresses both an essential protein and an inhibitor of RNA synthesis by shuffling modules via mRNA editing. EMBO J 10:3079–3085
Curran J, Marq J-B, Kolakofsky D (1992) The Sendai virus nonstructural C proteins specifically inhibit viral mRNA synthesis. Virology 189:647–656
Curran J, Homann H, Buchholz C, Rochat S, Neubert W, Kolakofsky D (1993) The hypervariable C-terminal tail of the Sendai paramyxovirus nucleocapsid protein in required for template function but not for RNA encapsidation. J Virol 67:4358–4364
Curran J, Pelet T, Kolakofsky D (1994) An acidic activation-like domain of the Sendai virus P protein is required for RNA synthesis and encapsidation. Virology 202:875–884
Curran J, Marq J-B, Kolakofsky D (1995) An N-terminal domain of the Sendai paramyxovirus P protein acts as a chaperone for the NP protein during the nascent chain assembly step of genome replication. J Virol 69:849–855
de Breyne S, Monney RS, Curran J (2004) Proteolytic processing and translation initiation: two independent mechanisms for the expression of the Sendai virus Y proteins. J Biol Chem 279:16571–16580
Delenda C, Hausmann S, Garcin D, Kolakofsky D (1997) Normal cellular replication of Sendai virus without the trans-frame, nonstructural V protein. Virology 228:55–62
Demirov DG, Freed EO (2004) Retrovirus budding. Virus Res 106:87–102
Didcock L, Young DF, Goodbourn S, Randall RE (1999a) Sendai virus and simian virus 5 block activation of interferon-responsive genes: importance for virus pathogenesis. J Virol 73:3125–3133
Didcock L, Young DF, Goodbourn S, Randall RE (1999b) The V protein of simian virus 5 inhibits interferon signalling by targeting STAT1 for proteasome-mediated degradation. J Virol 73:9928–9933
Drexler JF, Corman VM, Müller MA, Maganda GD, Vallo P, Binger T, Gloza-Rausch F, Rasche A, Yordanov S, Seebens A, Oppong S, Sarkodie Y, Pngombo C, Lukashev AN, Schmidt-Chanasit J, Stöcker A, Carneiro AJ, Erbar S, Maisner A, Fronhoffs F, Buettner R, Kalko EK, Kruppa T, Franke CR, Kallies R, Yandoko ER, Herrler G, Reusken C, Hassanin A, Krüger DH, Matthee S, Ulrich RF, Leroy EM, Drosten C (2012) Bats host major mammalian paramyxoviruses. Nat Commun 3:796. doi:10.1038/ncomms1796
Easterbrook JD, Kaplan JB, Glass GE, Watson J, Klein SL (2008) A survey of rodent-borne pathogens carried by wild-caught Norway rats: a potential threat to laboratory rodent colonies. Lab Anim 42:92–98
Egelman EH, Wu S-S, Amrein M, Portner A, Murti G (1989) The Sendai virus nucleocapsid exists in at least four different helical states. J Virol 63:2233–2243
Einberger H, Mertz R, Hofschneider PH, Neubert WJ (1990) Purification, renaturation, and reconstituted protein kinase activity of the Sendai virus large (L) protein: L protein phosphorylates the NP and P proteins in vitro. J Virol 64:4274–4280
Feller JA, Smallwood S, Horikami SM, Moyer SA (2000) Mutations in conserved domains IV and VI of the large (L) subunit of the Sendai virus RNA polymerase give an spectrum of defective RNA synthesis phenotypes. Virology 269:426–439
Finch JT, Gibbs AJ (1970) Observations on the suructure of the nucleocapsids of some paramyxoviruses. J Gen Virol 6:141–150

Fuerst TR, Niles EG, Studier FW, Moss B (1986) Eukaryotic transient-expression system based on recombinant vaccinia virus that synthesizes bacteriophage T7 RNA polymerase. Proc Natl Acad Sci USA 83:8122–8126

Fujii Y, Kiyotani K, Yashida T, Sakaguchi T (2001) Conserved and non-conserved regions in the Sendai virus genome: evolution of a gene possessing overlapping reading frames. Virus Genes 22:47–52

Fukuhara N, Huang C, Kiyotani K, Yoshida T, Sakaguchi T (2002) Mutational analysis of the Sendai virus V protein: importance of the conserved residues for Zn binding, virus pathogenesis and efficient RNA editing. Virology 299:172–178

Fukumi H, Nishikawa F, Kitayama T (1954) A pneumotropic virus from mice causing hemagglutination. Jpn J Med Sci Biol 7:345–363

Galinski MS, Troy RM, Banerjee AK (1992) RNA editing in the phosphoprotein gene of the human parainfluenza virus type 3. Virology 186:543–550

Garcin D, Pelet T, Calain P, Roux L, Curran J, Kolakofsky D (1995) A highly recombinogenic system for the recovery of infections Sendai paramyxovirus from cDNA: generation of a novel copy-back nondefective interfering virus. EMBO J 14:6087–6094

Garcin D, Latorre P, Kolakofsky D (1999) Sendai virus C proteins counteract the interferon-mediated induction of an antiviral state. J Virol 73:6559–6565

Gething MJ, White JM, Waterfield MD (1978) Purification of the fusion protein of Sendai virus: analysis of the NH_2-terminal sequence generated during precursor activation. Proc Natl Acad Sci USA 75:2737–2740

Goff PH, Gao Q, Palese P (2012) A majority of infectious Newcastle disease virus particles packages a single genome while a minority is multiploid. J Virol 86:10852–10856

Gosselin-Grenet AS, Mottet-Osman G, Roux L (2006) From assembly to virus particle budding: pertinence of the detergent resistant membranes. Virology 344:296–303

Gosselin-Grenet AS, Marq JB, Abrami L, Garcin D, Roux L (2007) Sendai virus budding in the course of an infection does not require Alix and VPS4A host factors. Virology 365:101–112

Gotoh B, Ogasawara T, Toyoda T, Inocencio NM, Hamaguchi M, Nagai Y (1990) An endoprotease homologous to the blood clotting factor X as a determinant of viral tropism in chick embryo. EMBO J 9:4189–4195

Gotoh B, Takeuchi K, Komatsu T, Yokoo J, Kimura Y, Kato A, Kurotani A, Nagai Y (1999) Knockout of the Sendai virus C genes eliminates the viral ability to prevent the interferon-α/β mediated responses. FEBS Lett 459:205–210

Gotoh B, Komatsu T, Takeuchi K, Yokoo J (2001) Paramyxovirus accessory proteins as interferon antagonists. Microbiol Immunol 45:787–800

Gotoh B, Komatsu T, Takeuchi K, Yokoo J (2002) Paramyxovirus strategies for evading the interferon response. Rev Med Virol 12:337–357

Gotoh B, Komatsu T, Takeuchi K, Yokoo J (2003a) The C-terminal half-fragment of the Sendai virus C protein prevents the gamma-activated factor from binding to a gamma-activated sequence site. Virology 316:29–40

Gotoh B, Takeuchi K, Komatsu T, Yokoo J (2003b) The STAT2 activation process is a crucial target of Sendai virus C protein for the blockade of alpha interferon signaling. J Virol 77:3360–3370

Grogan CC, Moyer SA (2001) Sendai virus wild-type and mutant C proteins show a direct correlation between L polymerase binding and inhibition of viral RNA synthesis. Virology 288:96–108

Gubbay O, Curran J, Kolakofsky D (2001) Sendai virus genome synthesis and assembly are coupled: a possible mechanism to promote viral RNA polymerase processivity. J Gen Virol 82:2895–2903

Gupta KC, Patwardhan S (1988) ACG, the initiator codon for Sendai virus C protein. J Biol Chem 263:8553–8556

Hamaguchi M, Yoshida T, Nishikawa K, Naruse H, Nagai Y (1983) Transcriptive complex of Newcastle disease virus. I. Both L and P proteins are required to constitute an active complex. Virology 128:105–117

Harrison MS, Sakaguchi T, Schmitt AP (2010) Paramyxovirus assembly and budding: building particles that transmit infections. Int J Biochem Cell Biol 42:1416

Hasan MK, Kato A, Shioda T, Sakai Y, Yu D, Nagai Y (1997) Creation of an infectious recombinant Sendai virus expressing the firefly luciferase gene from the 3′ proximal first locus. J Gen Virol 78:2813–2820

Hasan MK, Kato A, Muranaka M, Yamaguchi R, Sakai Y, Hatano I, Tashiro M, Nagai Y (2000) Versatility of the accessory C proteins of Sendai virus. Contribution to virus assembly as an additional role. J Virol 74:5619–5628

Heggeness MH, Scheid A, Choppin PW (1981) The relationship of conformational changes in the Sendai virus nucleocapsid to proteolytic cleavage of the NP polypeptide. Virology 114:555–562

Heggeness MH, Smith PR, Choppin PW (1982) In vitro assembly of the nonglycosylated membrane protein (M) of Sendai virus. Proc Natl Acad Sci U S A 79:6232–6236

Hendricks DD, Ono E, Seyer JM, Gupta KC (1993) Phosphorylation of the Sendai virus C proteins. Virology 197:471–474

Hewitt JA, Nermut MV (1977) A morphological study of the M-protein of Sendai virus. J Gen Virol 34:127–136

Heylbroeck C, Balachandran S, Servant MJ, DeLuca C, Barber GN, Lin R, Hiscott J (2000) The IRF-3 transcription factor mediates Sendai virus-induced apoptosis. J Virol 74:3781–3792

Hidaka Y, Kanda T, Iwasaki K, Nomoto A, Shioda T, Shibuta H (1984) Nucleotide sequence of Sendai virus genome region covering the entire M gene and the 3′ proximal 1013 nucleotides of the F gene. Nucleic Acids Res 12:7965–7973

Holmes DE, Moyer SA (2002) The phosphoprotein (P) binding site resides in the N terminus of the L polymerase subunit of Sendai virus. J Virol 76:3078–3083

Homma M, Ouchi M (1973) Trypsin action on the growth of Sendai virus in tissue culture cells. III. Structural difference of Sendai viruses grown in eggs and tissue culture cells. J Virol 12:1457–1465

Horikami SM, Curran J, Kolakofsky D, Moyer SA (1992) Comlpexes of Sendai virus NP-P and P-L proteins are required for defective interfering particle genome replication in vitro. J Virol 66:4901–4908

Horikami SM, Hector RE, Smallwood S, Moyer SA (1997) The Sendai virus C protein binds the L polymerase protein to inhibit viral RNA synthesis. Virology 235:261–270

Hornung V, Ellegast J, Kim S, Brzózka K, Jung A, Kato H, Poeck H, Akira S, Conzelmann KK, Schlee M, Endres S, Hartmann G (2006) 5′-Triphosphate RNA is the ligand for RIG-I. Science 314:994–997

Hosaka Y (1968) Isolation and structure of the nucleocapsid of HVJ. Virology 35:445–457

Hosaka Y, Kitano H, Ikeguchi S (1966) Studies on the pleomorphism of HVJ virions. Virology 29:205–221

Houben K, Marion D, Tarbouriech N, Ruigrok RWH, Blanchard L (2007) Interaction of the C-terminal domains of Sendai virus N and P proteins: comparison of polymerase-nucleocapsid interactions within the paramyxovirus family. J Virol 81:6807–6816

Howe C, Morgan C (1969) Interactions between Sendai virus and human erythrocytes. J Virol 3:70–81

Hu C-J, Gupta KC (2000) Functional significance of alternate phosphorylation in Sendai virus P protein. Virology 268:517–532

Hu C, Kato A, Bowman MC, Kiyotani K, Yoshida T, Moyer SA, Nagai Y, Gupta KC (1999) Role of primary constitutive phosphorylation of Sendai virus P and V proteins in viral replication and pathogenesis. Virology 263:195–208

Huang C, Kiyotani K, Fujii Y, Fukuhara N, Kato A, Nagai Y, Yoshida T, Sakaguchi T (2000) Involvement of the zinc-binding capacity of Sendai virus V protein in viral pathogenesis. J Virol 74:7834–7841

Hughes S, Mellstrom K, Kosik E, Tamanoi F, Brugge J (1984) Mutation of a termination codon affects *src* initiation. Mol Cell Biol 4:1738–1746

Huntley CC, De BP, Banerjee AK (1997) Phosphorylation of Sendai virus phosphoprotein by cellular protein kinase C ζ. J Biol Chem 272:16578–16584

Inoue M, Tokusumi Y, Ban H, Kanaya T, Tokusumi T, Nagai Y, Iida A, Hasegawa M (2003a) Nontransmissible virus-like particle formation by F-deficient Sendai virus is temperature sensitive and reduced by mutations in M and HN proteins. J Virol 77:3238–3246

Inoue M, Tokusumi Y, Ban H, Kanaya T, Shirakura M, Tokusumi T, Hirata T, Nagai Y, Iida A, Hasegawa M (2003b) A new type of Sendai virus vector deficient in the matrix gene has lost virus particle formation and gained extensive cell-to-cell spreading. J Virol 77:6419–6429

Irie T, Shimazu Y, Yoshida T, Sakaguchi T (2007) The YLDL sequence within Sendai virus M protein is critical for budding of virus-like particles and interacts with Alix/AIP1 independently of C protein. J Virol 81:2263–2273

Irie T, Nagata N, Yoshida T, Sakaguchi T (2008a) Paramyxovirus Sendai virus C proteins are essential for maintenance of negative-sense RNA genome in virus particles. Virology 374:495–505

Irie T, Nagata N, Yoshida T, Sakaguchi T (2008b) Recruitment of Alix/AIP1 to the plasma membrane by Sendai virus C protein facilitates budding of virus-like particles. Virology 371:108–120

Irie T, Inoue M, Sakaguchi T (2010) Significance of the YLDL motif in the M protein and Alix/AIP1 for Sendai virus budding in the context of virus infection. Virology 405:334–341

Irie T, Kiyotani K, Igarashi T, Yoshida A, Sakaguchi T (2012) Inhibition of interferon regulatory factor 3 activation by paramyxovirus V protein. J Virol 86:7136–7145

Irie T, Okamoto I, Yoshida A, Nagai Y, Sakaguchi T (2014) Sendai virus C proteins regulate viral genome and antigenome synthesis to dictate the negative genome polarity. J Virol 88:690–698

Ishida N, Homma M (1978) Sendai virus. Adv Virus Res 23:349–383

Ito T, Suzuki Y, Takada A, Kawamoto A, Otsuki K, Masuda H, Yamada M, Suzuki T, Kida H, Kawaoka Y (1997) Differences in sialic acid-galactose linkages in the chicken egg amnion and allantois influence human influenza virus receptor specificity and variant selection. J Virol 71:3357–3362

Iwata S, Schmidt AC, Titani K, Suzuki M, Kido H, Gotoh B, Hamaguchi M, Nagai Y (1994) Assignment of disulfide bridges in the fusion glycoprotein of Sendai virus. J Virol 68:3200–3206

Kato A, Sakai Y, Shioda T, Kondo T, Nakanishi M, Nagai Y (1996) Initiation of Sendai virus multiplication from transfected viral cDNA or RNA with negative or positive sense. Genes Cells 1:569–579

Kato A, Kiyotani K, Sakai Y, Yoshida T, Nagai Y (1997a) The paramyxovirus, Sendai virus, V protein encodes a luxury function required for viral pathogenesis. EMBO J 16:578–587

Kato A, Kiyotani K, Sakai Y, Yoshida T, Shioda T, Nagai Y (1997b) Importance of the cysteine-rich carboxyl-terminal half of V protein for Sendai virus pathogenesis. J Virol 71:7266–7272

Kato A, Kiyotani K, Hasan MK, Shioda T, Sakai Y, Yoshida T, Nagai Y (1999) Sendai virus gene start signals are not equivalent in reinitiation capacity: moderation at the F gene. J Virol 73:9237–9246

Kato A, Ohnishi Y, Kohase M, Saito S, Tashiro M, Nagai Y (2001) Y2, the smallest of the Sendai virus C proteins, is fully capable of both counteracting the antiviral action of interferons and inhibiting viral RNA synthesis. J Virol 75:3802–3810

Kato A, Ohnishi Y, Hishiyama M, Kohase M, Saito S, Tashiro M, Nagai Y (2002) The amino-terminal half of Sendai virus C protein is not responsible for either counteracting the antiviral action of interferons or down- regulating viral RNA synthesis. J Virol 76:7114–7124

Kato A, Cortese-Grogan C, Moyer SA, Sugahara F, Sakaguchi T, Kubota T, Otsuki N, Kohase M, Tashiro M, Nagai Y (2004) Characterization of the amino acid residues of sendai virus C protein that are critically involved in its interferon antagonism and RNA synthesis down-regulation. J Virol 78:7443–7454

Kato A, Kiyotani K, Kubota T, Yoshida T, Tashiro M, Nagai Y (2007) Importance of the anti-interferon capacity of Sendai virus C protein for pathogenicity in mice. J Virol 81: 3264–3271

Kido H, Yokogoshi Y, Sakai K, Tashiro M, Kishino Y, Fukutomi A, Katunuma N (1992) Isolation and characterization of a novel trypsin-like protease found in rat bronchiolar epithelial Clara cells: a possible activator of the viral fusion glycoprotein. J Biol Chem 267:13573–13579

Kido H, Okumura Y, Yamada H, Le TQ, Yano M (2007) Proteases essential for human influenza virus entry into cells and their inhibitors as potential therapeutic agents. Curr Pharm Des 13:405–414

Kimura Y, Ito Y, Shimokawa K, Nishiyama Y, Nagata I, Kitoh J (1975) Temperature-sensitive virus derived from BHK cells persistently infected with HVJ (Sendai virus). J Virol 15:55–63

Kiyotani K, Takao S, Sakaguchi T, Yoshida T (1990) Immediate protection of mice from lethal wild-type Sendai virus (HVJ) infections by a temperature-sensitive mutant, HVJpi, possessing homologous interfering capacity. Virology 177:65–74

Kiyotani K, Sakaguchi T, Kato A, Nagai Y, Yoshida T (2007) Paramyxovirus Sendai virus V protein counteracts innate virus clearance through IRF-3 activation, but not via interferon, in mice. Virology 359:82–91

Klenk HD, Garten W (1994) Host cell proteases controlling virus pathogenicity. Trends Microbiol 2:39–43

Kolakofsky D, Bruschi A (1975) Antigenomes in Sendai virions and Sendai virus-infected cells. Virology 66:185–191

Kolakofsky D, Boy de la Tour E, Bruschi A (1974) Self-annealing of Sendai virus RNA. J Virol 14:33–39

Kolakofsky D, Pelet T, Garcin D, Hausmann S, Curran J, Roux L (1998) Paramyxovirus RNA synthesis and the requirement for hexamer genome length: the rule of six revisited. J Virol 72:891–899

Kolokoltsov AA, Deniger D, Fleming EH, Roberts NJ Jr, Karpilow JM, Davey RA (2007) Small interfering RNA profiling reveals key role of clathrin-mediated endocytosis and early endosome formation for infection by respiratory syncytial virus. J Virol 81:7786–7800

Komatsu T, Takeuchi K, Yokoo J, Tanaka Y, Gotoh B (2000) Sendai virus blocks alpha interferon signaling to signal transducers and activators of transcription. J Virol 74:2477–2480

Komatsu T, Takeuchi K, Yokoo J, Gotoh B (2002) Sendai virus C protein impairs both phosphorylation and dephosphorylation processes of Stat1. FEBS Lett 511:139–144

Kondo T, Yoshida T, Miura N, Nakanishi M (1993) Temperature sensitive phenotype of a mutant Sendai virus strain is caused by its insufficient accumulation of the M protein. J Biol Chem 268:21924–21930

Koyama AH, Ogawa M, Kato A, Nagai Y, Adachi A (2001) Lack of apoptosis in Sendai virus-infected Hep-2 cells without participation of viral antiapoptosis gene. Microbes Infect 3:1115–1121

Koyama AH, Irie H, Kato A, Nagai Y, Adachi A (2003) Virus multiplicaiton and induction of apoptosis by Sendai virus: role of the C proteins. Microbes Infect 5:373–378

Kozak M (1984) Selection of initiation sites by eucaryotic ribosomes: effect of inserting AUG triplets upstream from the coding sequence for preproinsulin. Nucleic Acids Res 12:3873–3893

Kozak M (1987) Effects of intercistronic length on the efficiency of reinitiation by eucaryotic ribosomes. Mol Cell Biol 7:3438–3445

Kurotani A, Kiyotani K, Kato A, Shioda T, Sakai Y, Mizumoto K, Yoshida T, Nagai Y (1998) The Sendai virus, C proteins are categorically nonessential gene products but silencing their expression severely impairs viral replecation and pathogenesis. Genes Cells 3:111–124

Kuroya M, Ishida N, Shiratori T (1953) Newborn virus pneumonitis (type Sendai) II. Report: The isolation of a new virus possessing hemagglutinin activity. Yokohama Med Bull 4:217–233

Lamb RA, Choppin PW (1977a) The synthesis of Sendai virus polypeptides in infected cells. II. Instracellular distribution of polypeptides. Virology 81:371–381

Lamb RA, Choppin PW (1977b) The synthesis of Sendai virus polypeptides in infected cells. III. Phosphorylation of polypeptides. Virology 81:382–397

Lamb RA, Parks GD (2007) *Paramyxoviridae*: the viruses and their replication. In: Knipe DM, Howley PM (eds) Fields virology, vol 1, 5th edn. Lippincott, Williams & Wilkins, Philadelphia, pp 1449–1496

Lamb RA, Mahy BW, Choppin PW (1976) The synthesis of Sendai virus polypeptides in infected cells. Virology 69:116–131

Latorre P, Kolakofsky D, Curran J (1998) Sendai virus Y proteins are initiated by a ribosomal shunt. Mol Cell Biol 18:5021–5031

Leyrer S, Neubert WJ, Sedlmeier R (1998) Rapid and efficient recovery of Sendai virus from cDNA: factors influencing recombinant virus rescue. J Virol Methods 75:47–58

Li HO, Zhu Y-F, Asakawa M, Kuma H, Hirata T, Ueda Y, Lee Y-S, Fukumura M, Iida A, Kato T, Nagai Y, Hasegawa M (2000) A cytoplasmic RNA vector derived from non-transmissible Sendai virus with efficient gene transfer and expression. J Virol 74:6564–6569

Li M, Schmitt PT, Li Z, McCrory TS, He B, Schmitt AP (2009) Mumps virus matrix, fusion, and nucleocapsid proteins cooperate for efficient production of virus-like particles. J Virol 83:7261–7272

Liu C-C, Simonsen CC, Levinson AD (1984) Initiation of translation at internal AUG codons in mammalian cells. Nature (Lond) 309:82–85

Loney C, Mottet-Osman G, Roux L, Bhella D (2009) Paramyxovirus ultrastructure and genome packaging: cryo-electron tomography of Sendai virus. J Virol 83:8191–8197. doi:10.1128/JVI.00693-09

Luque LE, Russell CJ (2007) Spring-loaded heptad repeat residues regulate the expression and activation of paramyxovirus fusion protein. J Virol 81:3130–3141

Luque LE, Bridges OA, Mason JN, Boyd KL, Portner A, Russell CJ (2010) Residues in the heptad repeat a region of the fusion protein modulate the virulence of Sendai virus in mice. J Virol 84:810–821. doi:10.1128/JVI.01990-09

Lyles DS, Rupprecht CE (2007) Rhabdoviridae. In: Knipe DM, Howley PM (eds) Fields virology, vol 1, 5th edn. Lippincott, Williams & Wilkins, Philadelphia, pp 1363–1408

Markwell MA, Fox CF (1980) Protein–protein interactions within paramyxoviruses identified by native disulfide bonding or reversible chemical cross-linking. J Virol 33:152–166

Markwell MAK, Paulson JC (1980) Sendai virus utilizes specific sialyloligosaccharides as host cell receptor determinants. Proc Natl Acad Sci USA 77:5693–5697

Markwell MA, Portner A, Schwartz AL (1985) An alternative route of infection for viruses: entry by means of the asialoglycoprotein receptor of a Sendai virus mutant lacking its attachment protein. Proc Natl Acad Sci USA 82:978–982

Marq J-B, Brini A, Kolakofsky D, Garcin D (2007) Targeting of the Sendai virus C protein to the plasma membrane via a peptide-only membrane anchor. J Virol 81:3187–3197

Matsuoka Y, Curran J, Pelet T, Kolakofsky D, Ray R, Compans RW (1991) The P gene of human parainfluenza virus type 1 encodes P and C proteins but not a cysteine-rich V protein. J Virol 65:3406–3410

Miyauchi K, Kim Y, Latinovic O, Morozov V, Melikyan GB (2009) HIV enters cells via endocytosis and dynamin-dependent fusion with endosomes. Cell 137:433–444

Morgan C, Howe C (1968) Structure and development of viruses as observed in the electron microscope. IX. Entry of parainfluenza I (Sendai) virus. J Virol 2:1122–1132

Mottet G, Roux L (1989) Budding efficiency of Sendai virus nucleocapsids: influence of size and ends of the RNA. Virus Res 14:175–187

Mottet-Osman G, Iseni F, Pelet T, Wiznerowics M, Garcin D, Roux L (2007) Suppression of the Sendai virus M protein through a novel short interfering RNA approach inhibits viral particle production but does not affect viral RNA analysis. J Virol 81:2861–2868

Moyer SA, Baker SC, Lessard JL (1986) Tublin: a factor necessary for the synthesis of both Sendai virus and vesicular stomatitis virus RNAs. Proc Natl Acad Sci USA 83:5405–5409

Murphy AM, Grdzelishvili VZ (2009) Identification of Sendai virus L protein amino acid residues affecting viral mRNA cap methylation. J Virol 83:1669–1681. doi:10.1128/JVI.01438-08

Myers TM, Moyer SA (1997) An amino-terminal domain of the Sendai virus nucleocapsid protein is required for template function. J Virol 71:918–924

Nagai Y (1993) Protease-dependent virus tropism and pathogenicity. Trends Microbiol 1:81–87

Nagai Y (1995) Virus activation by host proteinases. A pivotal role in the spread of infection, tissue tropism and pathogenicity. Microbiol Immunol 39:1–9

Nagai Y (1999) Paramyxovirus replication and pathogenesis. Reverse genetics transforms understanding. Rev Med Virol 9:83–99

Nagai Y, Kato A (1999) Paramyxovirus reverse genetics is coming of age. Microbiol Immunol 43:613–624

Nagai Y, Kato A (2004) Accessory genes of the *Paramyxoviridae*, a large family of nonsegmented negative-strand and RNA viruses, as a focus of active investigation by reverse genetics. Curr Top Microbiol Immunol 283:197–248

Nagai Y, Yoshida T (1984) Viral pathogenesis: mechanism of acute and persistent infections with paramyxoviruses. Nagoya J Med Sci 46:1–17

Nagai Y, Ogura H, Klenk H-D (1976a) Studies on the assembly of the envelope of Newcastle disease virus. Virology 69:523–538

Nagai Y, Klenk H-D, Rott R (1976b) Proteolytic cleavage of the viral glycoproteins and its significance for the virulence of Newcastle disease virus. Virology 72:494–508

Nagai Y, Hamaguchi M, Toyoda T, Yoshida T (1983a) The uncoating of paramyxoviruses may not require a low pH mediated step. Virology 130:263–268

Nagai Y, Yoshida T, Hamaguchi M, Nagura H, Hasegawa H, Yoshimura S, Watanabe K (1983b) Subcellular location of the major protein antigens of paramyxoviruses revealed by immunoperoxidase cytochemistry. Microbiol Immunol 27:531–545

Nagai Y, Inoue M, Iida A, Zhu Y-F, Hasegawa M, Kato A, Matano T (2007) Sendai virus engineering: from reverse genetics to vector development. In: Hefferon KL (ed) Virus expression vectors. Transworld Research Network, Kerala, pp 123–146

Nagai Y, Takakura A, Irie T, Yonemitsu Y, Gotoh B (2011) Sendai virus evolution from mouse pathogen to a state-of-the-art tool in virus research and biotechnology. In: Samal SK (ed) The biology of paramyxoviruses. Caister Academic, Norfolk

Nagata I, Kimura Y, Ito Y, Tanaka T (1972) Temperature-sensitive phenomenon of viral maturation observed in BHK cells persistently infected with HVJ. Virology 49:453–461

Nishimura K, Segawa H, Goto T, Morishita M, Masago A, Takahashi H, Ohmiya Y, Sakaguchi T, Asada M, Imamura T, Shimotono K, Takayama K, Yoshida T, Nakanishi M (2007) Persistent and stable gene expression by a cytoplasmic RNA replicon based on a noncytopathic variant Sendai virus. J Biol Chem 282:27383–27391

Nishio M, Nagata A, Tsurudome M, Ito M, Kawano M, Komada H, Ito Y (2004) Recombinant Sendai viruses with L1618V mutation in their L polymerase protein establish persistent infection, but not temperature sensitivity. Virology 329:289–301

Ogasawara T, Gotoh B, Suzuki H, Asaka J, Shimokata K, Rott R, Nagai Y (1992) Expression of factor X and its significance for the determination of paramyxovirus tropism in the chick embryo. EMBO J 11:467–472

Ogino T, Kobayashi M, Iwama M, Mizumoto K (2005) Sendai virus RNA-dependent RNA polymerase L protein catalyzes cap methylation of virus-specific mRNA. J Biol Chem 280:4429–4435

Ohnishi Y, Shioda T, Nakayama K, Iwata S, Gotoh B, Hamaguchi M, Nagai Y (1995) A furin-defective cell line is able to process correctly the gp160 of human immunodeficiency virus type 1. J Virol 68:4075–4079

Palese P, Zheng H, Engelhardt OG, Pleschka S, García-Sastre A (1996) Negative-strand RNA viruses: genetic engineering and applications. Proc Natl Acad Sci USA 93:11354–11358

Pennington TH, Pringle CR (1978) Negative strand viruses in enucleate cells. In: Mahy BWJ, Barry RD (eds) Negative strand viruses and the host cell. Academic, New York

Peters K, Chattopadhyay S, Sen GC (2008) IRF-3 activation by Sendai virus infection is required for cellular apoptosis and avoidance of persistence. J Virol 82:3500–3508

Poch O, Blumberg BM, Bougueleret L, Tordo N (1990) Sequence comparison of five polymerases (L proteins) of unsegmented negative-strand RNA viruses: theoretical assignment of functional domains. J Gen Virol 71:1153–1162

Portner A, Murti KG (1986) Localization of P, NP, and M proteins on Sendai virus nucleocapsid using immunogold labeling. Virology 150:469–478

Portner A, Gupta KC, Seyer JM, Beachey EH, Kingsbury DW (1986) Localization and characterization of Sendai virus nonstructural C and C' proteins by antibodies against synthetic peptides. Virus Res 6:109–121

Portner A, Murti KG, Morgan EM, Kingsbury DW (1988) Antibodies against Sendai virus L protein: distribution of the protein in nucleocapsids revealed by immunoelectron microscopy. Virology 163:236–239

Qanungo KR, Shaji D, Mathur M, Banerjee AK (2004) Two RNA polymerase complexes from vesicular stomatitis virus-infected cells that carry out transcription and replication of genome RNA. Proc Natl Acad Sci USA 101:5952

Radoshitzky SR, Dong L, Chi X, Clester JC, Retterer C, Spurgers K, Kuhn JH, Sandwick S, Ruthel G, Kota K, Bolts D, Warren T, Kranzusch PJ, Whelan SPJ, Bavari S (2010) Infectious Lassa virus, but not filoviruses, is restricted by BST-2/tetherin. J Virol 84:10569–10580. doi:10.1128/JVI.00103-10

Randall RE, Goodbourn S (2008) Interferons and viruses: an interplay between induction, signalling, antiviral responses and virus countermeasures. J Gen Virol 89:1–47

Randall RE, Russell WC (1991) Paramyxovirus persistence: consequences for host and virus. In: Kingsbury DW (ed) The paramyxoviruses. Plenum, New York, pp 299–321

Re GG (1991) Deletion mutants of paramyxoviruses. In: Kingsbury DW (ed) The paramyxoviruses. Plenum, New York, pp 275–298

Re GG, Kingsbury DW (1986) Nucleotide sequences that affect replicative and transcriptional efficiencies of Sendai virus deletion mutants. J Virol 58:578–582

Rima BK, Martin SJ (1976) Persistent infection of tissue culture cells by RNA viruses. Med Microbiol Immunol 162:89–118

Roux L, Holland JJ (1979) Role of defective interfering particles of Sendai virus in persistent infections. Virology 93:91–103

Roux L, Holland JJ (1980) Viral genome synthesis in BHK-21 cells persistently infected with Sendai virus. Virology 100:53–64

Ryan KW, Morgan EM, Portner A (1991) Two noncontiguous regions of Sendai virus P protein combine to form a single nucleocapsid binding domain. Virology 180:126–134

Sakaguchi T, Toyoda T, Gotoh B, Inocencio NM, Kuma K, Miyata T, Nagai Y (1989) Newcastle disease virus evolution. I. Multiple lineages defined by sequence variability of the hemagglutinin-neuraminidase gene. Virology 169:260–272

Sakaguchi T, Kiyotani K, Kato A, Asakawa M, Fujii Y, Nagai Y, Yoshida T (1997) Phosphorylation of the Sendai virus M protein is not essential for virus replication either in vitro or in vivo. Virology 235:360–366

Sakaguchi T, Uchiyama T, Huang C, Fukuhara N, Kiyotani K, Nagai Y, Yoshida T (2002) Alteration of Sendai virus morphogenesis and nucleocapsid incorporation due to mutation of cycteine residues of the matrix protein. J Virol 76:1682–1690

Sakaguchi T, Kato A, Sugahara F, Shimazu Y, Inoue M, Kiyotani K, Nagai Y, Yoshida T (2005) AIP1/Alix is a binding partner of Sendai virus C protein and facilitates virus budding. J Virol 79:8933–8941

Sakaguchi T, Kato A, Kiyotani K, Yoshida T, Nagai Y (2008) Studies on the paramyxovirus accessory genes by reverse genetics in the Sendai virus–mouse system. Proc Jpn Acad Ser B Phys Biol Sci 84:439–451

Sanderson CM, Avalos R, Kundu A, Nayak DP (1995) Interaction of Sendai virus F, HN, and M proteins with host cytoskeletal and lipid components in Sendai virus-infected BHK cells. Virology 209:701–707

Scheid A, Choppin PW (1974) Identification of biological activities of paramyxovirus glycoproteins. Activation of cell fusion, hemolysis, and infectivity of proteolytic cleavage of an inactive precursor protein of Sendai virus. Virology 57:475–490

Schmitt AP, Leser GP, Morita E, Sundquist WI, Lamb RA (2005) Evidence for a new viral late domain core sequence, FPIV, necessary for budding of a paramyxovirus. J Virol 79:2988–2997

Schnell MJ, Mebatsion T, Conzelmann K-K (1994) Infectious rabies virus from cloned cDNA. EMBO J 13:4195–4203

Segawa H, Yamashita T, Kawakita M, Taira H (2000) Functional analysis of the individual oligosaccharide chains of Sendai virus fusion protein. J Biochem 128:65–72

Segawa H, Inakawa A, Yamashita T, Taira H (2003) Functional analysis of individual oligosaccharide chains of Sendai virus hemagglutinin-neuraminidase protein. Biosci Biotechnol Biochem 67:592–598

Shimazu Y, Takao SI, Irie T, Kiyotani K, Yoshida T, Sakaguchi T (2008) Contribution of the leader sequence to homologous viral interference among Sendai virus strains. Virology 372:64–71

Shimizu K, Ishida N (1975) The smallest protein of Sendai virus: its candidate function of binding nucleocapsid to envelope. Virology 67:427–437

Shimizu YK, Shimizu K, Ishida N, Homma M (1976) On the study of Sendai virus hemolysis. II. Morphological study of envelope fusion and hemolysis. Virology 71:48–60

Shioda T, Hidaka Y, Kanda T, Shibuta H, Nomoto A, Iwasaki K (1983) Sequence of 3,687 nt from the 3′ end of Sendai virus genome RNA and the predicted amino acid sequences of viral NP, P and C proteins. Nucleic Acids Res 11:7317–7330

Shioda T, Iwasaki K, Shibuta H (1986) Determination of the complete nucleotide sequence of the Sendai virus genome RNA and the predicted amino acid sequence of the F, HN, and L proteins. Nucleic Acids Res 14:1545–1563

Shirogane Y, Takeda T, Iwasaki M, Ishiguro N, Takeuchi H, Nakatsu Y, Tahara M, Kikuta H, Yanagi Y (2008) Efficient multiplication of human metapneumovirus in Vero cells expressing the transmembrane serine protease TMPRSS2. J Virol 82:8942–8946

Smallwood S, Ryan KW, Moyer SA (1994) Deletion analysis defines a carboxyl-proximal region of Sendai virus P protein that binds to the polymerase L protein. Virology 202:154–163

Smallwood S, Hövel T, Neubert WJ, Moyer SA (2002a) Different substitutions at conserved amino acids in domains II and III in the Sendai L RNA polymerase protein inactivate viral RNA synthesis. Virology 304:135–145

Smallwood S, Çevik B, Moyer SA (2002b) Intragenic complementation and oligomerization of the L subunit of the Sendai virus RNA polymerase. Virology 304:235–245

Sriwilaijaroen N, Kondo S, Yagi H, Wilairat P, Hiramatsu H, Ito M, Ito Y, Kato K, Suzuki Y (2009) Analysis of *N*-glycans in embryonated chicken egg choriollantoic and amniotic cells responsible for binding and adaptation of human and avian influenza viruses. Glycoconj J 26:433–443

Sugahara F, Uchiyama T, Watanabe H, Shimazu Y, Kuwayama M, Fujii Y, Kiyotani K, Adachi A, Kohno N, Yoshida T, Sakaguchi T (2004) Paramyxovirus Sendai virus-like particle formation by expression of multiple viral proteins and acceleration of its release by C protein. Virology 325:1–10

Suzuki Y, Suzuki T, Matsumoto M (1983) Isolation and characterization of receptor sialoglycoprotein for hemagglutinating virus of Japan (Sendai virus) from bovine erythrocyte membrane. J Biochem (Tokyo) 93:1621–1633

Suzuki Y, Suzuki T, Matsunaga M, Matsumoto M (1985) Gangliosides as paramyxovirus receptor. Structural requirement of sialo-oligosaccharides in receptors for hemagglutinating virus of Japan (Sendai virus) and Newcastle disease virus. J Biochem (Tokyo) 97:1189–1199

Suzuki H, Harada A, Hayashi Y, Wada K, Asaka J, Gotoh B, Ogasaweara T, Nagai Y (1991) Primary structure of the virus activating protease from chick embryo; its identity with the blood clotting factor Xa. FEBS Lett 283:281–285

Takimoto T, Bousse T, Coronel EC, Scroggs RA, Portner A (1998) Cytoplasmic domain of Sendai virus HN protein contains a specific sequence required for its incorporation into virions. J Virol 72:9747–9754

Tamura T, Yamashita T, Segawa H, Taira H (2002) *N*-linked oligosaccharide chains of Sendai virus fusion protein determine the interaction with endoplasmic reticulum molecular chaperones. FEBS Lett 513:153–158

Tapparel C, Hausmann S, Pelet T, Curran J, Kolakofsky D, Roux L (1997) Inhibition of Sendai virus genome replication due to promoter-increased selectivity: a possible role for the accessory C proteins. J Virol 71:9588–9599

Tapparel C, Maurice D, Roux L (1998) The activity of Sendai virus genomic and antigenomic promoters requires a second element past the leader template regions: a motif, $(GNNNNN)_3$, is essential for replication. J Virol 72:3117–3128

Tarbouriech N, Curran J, Ebel C, Ruigrok RWH, Burmeister WP (2000a) On the domain structure and the polymerization state of the Sendai virus P protein. Virology 266:99–109

Tarbouriech N, Curran J, Ruigrok RWH, Burmeister WP (2000b) Tetrameric coiled coil domain of Sendai virus phosphoprotein. Nat Struct Biol 7:777–781

Tashiro M, Yokogoshi Y, Tobita K, Seto JT, Rott R, Kido H (1992) Tryptase Clara, an activating protease for Sendai virus in rat lungs, is involved in pneumopathogenicity. J Virol 66:7211–7216

Toyoda T, Sakaguchi T, Imai K, Inocencio NM, Gotoh B, Hamaguchi M, Nagai Y (1987) Structural comparison of the cleavage-activation site of the fusion glycoprotein between virulent and avirulent strains of Newcastle disease virus. Virology 158:242–247

Toyoda T, Sakaguchi T, Hirota H, Gotoh B, Kuma K, Miyata T, Nagai Y (1989) Newcastle disease virus evolution. II. Lack of gene recombination in generating virulent and avirulent strains. Virology 169:273–282

Tozawa H, Watanabe M, Ishida N (1973) Structural components of Sendai virus. Serological and physicochemical characterization of hemagglutinin subunit associated with neuraminidase activity. Virology 55:242–253

Vidal S, Curran J, Kolakofsky D (1990a) Editing of the Sendai virus P/C mRNA by G insertion occurs during mRNA synthesis via a virus-encoded activity. J Virol 64:239–246

Vidal S, Curran J, Kolakofsky D (1990b) A stuttering model for paramyxovirus P mRNA editing. EMBO J 9:2017–2022

Watanabe R, Leser GP, Lamb RA (2011) Influenza virus is not restricted by tetherin whereas influenza VLP production is restricted by teterin. Virology 417:50–56

Whelan SP, Wertz GW (2002) Transcription and replication initiate at separate sites on the vesicular stomatitis virus genome. Proc Natl Acad Sci USA 99:9178–9183

Whelan SPJ, Barr JN, Wertz GW (2004) Transcription and replication of nonsegmented negative-strand RNA viruses. Curr Top Microbiol Immunol 283:61–119

Wiegand MA, Bossow S, Schlecht S, Neubert WJ (2007) De novo synthesis of N and P proteins as a key step in Sendai gene expression. J Virol 81:13835–13844. doi:10.1128/JVI.00914-07

Yamada H, Hayata S, Omata-Yamada T, Taira H, Mizumoto K, Iwasaki K (1990) Association of the Sendai virus C protein with nucleocapsids. Arch Virol 113:245–253

Yoshida T, Nagai Y, Yoshii S, Maeno K, Matsumoto T, Hoshino M (1976) Membrane (M) protein of HVJ (Sendai virus): its role in virus assembly. Virology 71:143–161

Yoshida T, Nagai Y, Maeno K, Iinuma M, Hamaguchi M, Matsumoto T (1979) Studies on the role of M protein in virus assembly using a TS mutant of HVJ (Sendai virus). Virology 92:139–154

Yoshida T, Hamaguchi M, Naruse H, Nagai Y (1982) Persistent infection by a temperature-sensitive mutant isolated from a Sendai virus (HVJ) carrier culture: its initiation and maintenance without aid of defective interfering particles. Virology 120:329–339

Yoshima H, Nakanishi M, Okada Y, Kobata A (1981) Carbohydrate structures of HVJ (Sendai virus) glycoproteins. J Biol Chem 256:5355–5361

Young DF, Didcock L, Goodbourn S, Randall RE (2000) Paramyxoviridae use distinct virus-specific mechanisms to circumvent the interferon response. Virology 269:383–390

Youngner JS, Preble OT (1980) Viral persistence: evolution of viral populations. In: Frankel-Conrat H, Wagner RR (eds) Comprehensive virology, vol 16. Plenum, New York/London, pp 73–135

Chapter 3
Concept and Technology Underlying Sendai Virus (SeV) Vector Development

Akihiro Iida and Makoto Inoue

Abstract Sendai virus (SeV) reverse genetics, hatched out of pure academic need, is developing into the leading-edge biotechnology of a nonintegrating, cytoplasmic RNA vector. Diverse applications of the SeV vector are now in the pipeline and are poised for clinical testing. Here, we describe the conceptual novelty of the SeV vector and the technology used to accommodate a foreign gene of interest into the full-length SeV genome. This technology has advanced further to the generation of non-transmissible safer vector versions. Moreover, the incorporation of predetermined mutations responsible for particular phenotypes into the SeV backbone was actively conducted to produce special versions for its use in respective special settings.

3.1 Introduction

Upon the establishment of SeV reverse genetics, we had the idea of developing a SeV vector system with a completely new concept, that is, a cytoplasmic RNA replicon vector that is safer and more useful than the then-existing viral vectors. Full-length SeV cDNA was initially used as the vector backbone. Technology then advanced to the creation of the second-generation SeV vectors having various SeV essential genes deleted, with the primary aim of rendering the vectors nontransmissible to increase their safety. Special engineering such as the introduction of predetermined phenotype-specific mutations expanded the usability of the SeV vector for diverse medical and biological purposes. In this chapter, we outline the history of SeV vector development from primitive to advanced versions, along with discussions on the potential advantages and utilities of the SeV vector.

A. Iida (✉) • M. Inoue
DNAVEC Corporation, 6 Ohkubo, Tsukuba-shi, Ibaraki, 300-2611, Japan
e-mail: akihiro.iida@384.jp

Y. Nagai (ed.), *Sendai Virus Vector: Advantages and Applications*,
DOI 10.1007/978-4-431-54556-9_3,

3.2 Feasibility of Using SeV as a Novel Class of Expression Vector: An Overview

3.2.1 No Potential Pathogenicity for Humans

Safety is one of the great concerns associated with medical use of any viral vectors. No link between SeV and any human diseases has been reported for the nearly six decades since the discovery of the virus. It has been used in laboratories for a long period of time and has caused no accidental infections in humans. In a clinical study, intranasal inoculation of SeV in adult volunteers produced no harmful outcomes (Slobod et al. 2004). Similarly, intramuscular inoculations as great as 5×10^9 CIU (cell infectious units)/60 kg produced no adverse events in a clinical trial of gene therapy for peripheral arterial disease (PAD) using the *F* gene-deleted SeV vector expressing human fibroblast growth factor-2 (FGF-2) (Yonemitsu et al. 2013; see Chap. 8).

3.2.2 Nonintegrating Nature

One of the most frequently used viral vectors to date in human gene therapy was derived from the genus *Gammaretrovirus* in the family *Retroviridae* (Thomas et al. 2003; Verma and Weitzman 2005). Integration of the vector genome (i.e., its DNA copied via reverse transcription) into chromosomes of target cells has been a precondition for *trans*-gene expression from these retroviral vectors, raising fear or safety concerns such as proto-oncogene activation and other disorders in cellular gene expression, which are comprehensively regarded as a genotoxicity concern. Clinical trials of the treatment of patients with X-linked severe combined immunodeficiency (X-SCID) were carried out in Paris and London from 2001, in which it was found that hematopoietic stem cells had been transduced with a retroviral vector carrying the gene encoding the common gamma chain of the interleukin-2 receptor (IL2RG). This trial resulted in the development of T-cell active lymphoblastic leukemia in 5 of a total of 20 patients treated, which appeared to be caused by the activation of the proto-oncogene LIM-domain-only 2 (LMO2) in 4 patients and cyclin D2 (CCND2) in 1 patient following vector insertion to a site(s) proximal to these proto-oncogenes in the patients' chromosomes (Hacein-Bey-Abina et al. 2003, 2008).

Adenoviral and adeno-associated viral vectors are also frequently used. Their chromosomal integration is less frequent than that of retroviral vectors. However, these DNA viruses have a nuclear phase in their life cycle, which does not completely eliminate the possibility of the recombination of these vectors with chromosomes that may cause some form of genotoxicity.

In contrast, with the exception of the *Bornaviridae*, SeV and all other nonsegmented negative strand RNA viruses (the *Mononegavirales*) have no nuclear phase throughout their life cycles, which are completed entirely in the cytoplasm in an

RNA replicon form (reviewed in Conzelmann 1998; Lamb and Parks 2007; Nagai et al. 2011). Therefore, these viruses replicate in enucleated cells or in the presence of the inhibitors of DNA-dependent RNA polymerase such as actinomycin D and α-amanitin (Armeanu et al. 2003, see Chap. 2). In theory, these features of the *Mononegavirales* eliminate the possibility of genotoxicity caused by integration into and any other interactions with chromosomes, representing the first-line advantage of SeV and related viruses in vector generation.

3.2.3 Lack of Homologous/Heterologous Recombination and Genome Mixing

Systematic phylogenetic analyses suggested that virtually no recombination events occurred in the evolutionary process of the *Mononegavirales* (Chare et al. 2003). No homologous or heterologous recombination has been reported over a long period of SeV research, eliminating the possibility of the emergence of unforeseen recombinant viruses. This situation may be partly explained by the fact that the SeV RNA genome is tightly bound with N proteins to form a ribonucleoprotein (RNP) complex and does not become naked RNA throughout the life cycle; hence, it has minimal opportunities for base pairing. The SeV polymerase can sometimes jump from one position to another during replication and continue to copy a distal region of the genome template to yield a deletion-type defective interfering (DI) particle or copy back the nascent chain to yield a copy-back DI particle (Re et al. 1985). This change may leave open the possibility that homologous recombination may occur, albeit with low efficiency, when two distinct RNP strands are replicating in a single cell. However, there have been no reports to date that support this possibility,

An additional advantage of using SeV and the *Mononegavirales* as expression vectors is that they possess a single, nonsegmented RNA genome and, hence, do not undergo the gene reassortment seen in RNA viruses with segmented genomes such as orthomyxo-, arena-, bunya-, and reoviruses (reviewed in Simon-Loriere and Holmes 2011).

3.2.4 Very High Performance in Foreign Gene Expression

After entering host cells, SeV initiates transcription and replication in the cytoplasm by its RNA-dependent RNA polymerase. As the SeV accessory C protein encodes an anti-apoptosis function (see Chap. 2), its cytopathic effects are not very extensive, allowing a fairly long time span of infected cell survival. Because the amplification of genomic RNA and mRNAs is vigorous, SeV replicates to high titers along with a high level of viral protein expression in cells. Consequently, the continued, robust expression of inserted foreign genes is expected (see Sect. 3.3.3). However, the robust expression of *trans*-genes is not always advantageous and can sometimes be hazardous. In this context, downregulating the expression of foreign genes to various extents is possible by selecting the site of their insertion in the SeV genome (see Sect. 3.5.2).

Because SeV reaches as high as 10^{10} CIU (cell infectious units)/mL in the allantoic fluid of embryonated chicken eggs and 10^{8-9} CIU/mL in the culture supernatant of cell lines or primary cells, the production of vector virus stocks with high titers is guaranteed, which represents a critically important justification for its practical use.

3.2.5 Very Broad Target Cell Range

Conventional gammaretrovirus vectors frequently derived from Moloney murine leukemia virus cannot infect nondividing cells such as neurons and muscle cells. The adenovirus hardly infects blood cells and hardly enters through the apical surface of epithelial cells. The main site of virus entry in epithelial tissues appears to be the basolateral surface because of the polarized localization on the basolateral surface of the specific receptor for the fiber proteins, CAR (the coxsackie virus and adenovirus receptor) (Walters et al. 1999). Therefore, gene transfer through epithelial tissues by the adenovirus vector is not easy.

SeV uses as its receptor the terminal sialic acids of glycosylated molecules ubiquitously present on the apical surface of epithelial cells as well as other cell surfaces. SeV can readily infect both dividing and nondividing cells and exhibits a very broad target cell range. Thus, a wide variety of diseases can potentially be the therapeutic targets of the SeV vector. Importantly, SeV tropism is restricted because the tryptic endoprotease required for activating the viral fusion protein is expressed in limited tissue types only (Nagai 1993). Thus, SeV vector stocks have to be activated before use by cleaving the inactive precursor fusion protein with a low concentration of trypsin (Homma and Ohuchi 1973, see Chap. 2; also see Sect. 3.5.5). This is another key to the safety of the SeV vector because the progeny produced from initially infected cells does not gain infectivity in the absence of such enzymes.

3.2.6 Additional Advantageous Aspects

In addition to the advantages described of the characteristics of SeV as a gene transfer vector, the following properties of SeV are also listed as additional advantages. SeV can attach to target cells and transfer its genome into these cells instantly, within seconds (Masaki et al. 2001; Ikeda et al. 2002). This ability is of particular merit in a situation where, for example, stopping the bloodstream is required during gene delivery. SeV is also able to traverse the mucous layer of the nasal cavity and airway *in vivo*, readily accessing the epithelial cell surface (Yonemitsu et al. 2000; Griesenbach et al. 2002).

3.2.7 Antigenic Cross-Reactivity with Human Parainfluenza Virus Type 1 (hPIV-1)

SeV and human parainfluenza virus type 1 (hPIV-1) show limited antigenic cross-reactivity with each other (Komada et al. 1992). Human populations are believed to be often exposed to hPIV-1, many of whom are expected to possess neutralizing antibodies to SeV, thus raising a doubt about the efficacy of human use of SeV vectors in general. A large-scale surveillance for SeV-neutralizing antibodies in humans was made (Hara et al. 2011). In addition, repeated use of SeV vectors in monkeys was attempted (Kurihara et al. 2012). Overall, the data so far suggested the feasibility of using SeV vectors in humans (For details, see Chap. 5). However, a clear picture on this issue will emerge only after clinical trials in various settings.

3.3 First-Generation SeV Vector with the Full-Length Wild-Type RNA Genome

3.3.1 Protocol for Foreign Gene Insertion into the cDNA of the SeV Genome

The first generation of SeV vectors carrying a foreign gene in the full-length SeV genome of 15,384 nucleotides (nt) was created according to the following protocol (Hasan et al. 1997). An 18-nt-long sequence containing a unique *Not*I site was introduced in the 5′- (in the positive sense) noncoding region located 41 nt downstream of the S (start) signal for *N*-gene transcription in the SeV cDNA backbone. A *NotI-tagged* polymerase chain reaction (PCR)-amplified product was inserted into the *Not*I site. The S signal originally used for *N*-gene transcription was then expected to act as the S signal for expression of the inserted gene. Because transcription of the inserted gene had to be terminated so that transcription of the next gene could be initiated, the antisense primer was designed just before the *Not*I tag to possess synthetic E (end, termination) and S (reinitiation) signals connected with a three-nucleotide intergenic sequence (I). The SeV transcription unit consisted of I + S+ (viral gene or foreign gene) + E (Fig. 3.1). All manipulations were performed so that the length of the final product would be an even multiple of six nucleotides (the rule of six: Calain and Roux 1993; Kolakofsky et al. 1998; Vulliemoz and Roux 2001; also see Chap. 2). The foreign gene was placed between the 3′-leader (*le*) sequence and the *N* gene; hence, this insertion was defined as *le/N* (also see Sect. 3.5.2). Other groups also generated recombinant SeV vectors by inserting a PCR-amplified foreign gene in a similar cassette-like fashion (Bitzer et al. 2003a, b; Nishimura et al. 2007). There were two replication promoter elements, PrE-I and PrE-II, each consisting of 18 (6×3) nucleotides. They were separated by 60 nucleotides so that they were juxtaposed on the same surface of the helical RNP with a turn of 6×13 nucleotides. This positioning was shown to be critical for recognition and replication

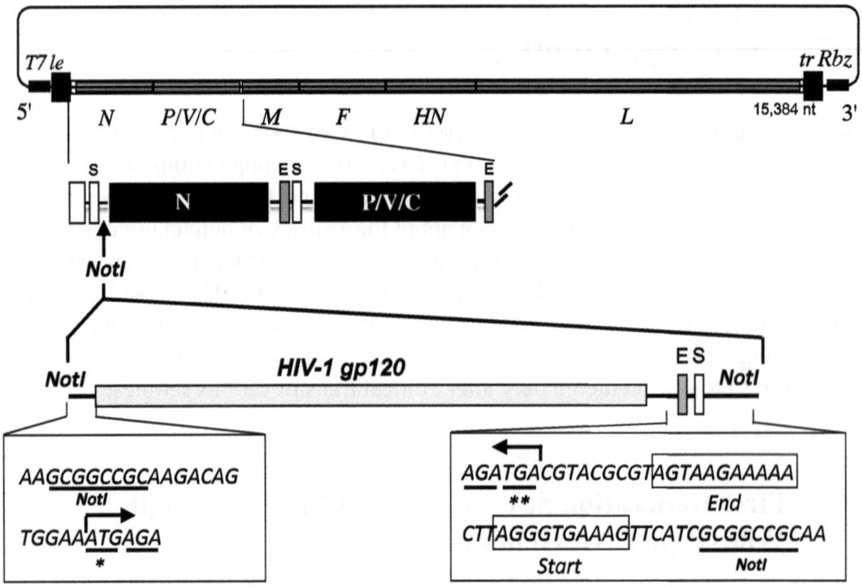

Fig. 3.1 Construction of a plasmid carrying a foreign gene in antigenomic SeV cDNA. An 18-nucleotide sequence with a unique *Not*I site was introduced between the start signal and open reading frame of the *N* gene. A PCR-amplified fragment carrying a foreign gene (e.g., HIV1 gp 120) and *Not*I at both ends was then inserted into the *Not*I site of the parental SeV cDNA. This plasmid was used to generate the recombinant SeV vector together with other supportive plasmids

by the RNA polymerase (P and L proteins) (Tapparel et al. 1998: reviewed in Nagai et al. 2011; also see Chap. 2). The first transcription start signal was within the spacer. All manipulations were conducted without disturbing the correct positioning of PrE-I and PrE-II (see Chap. 2).

3.3.2 Recovery of Recombinant SeV Using the T7 Promoter

The procedure to recover SeV vectors with the wild-type (full-length) backbone followed the protocol to recover wild-type SeV (Kato et al. 1996; reviewed in Nagai and Kato 1999; also see Chap. 2). Briefly, four plasmids carrying the cDNAs of the SeV antigenome and the *N*, *P*, and *L* genes, respectively, under the control of the T7 promoter from the T7 bacteriophage were transfected into LLC-MK2 cell line of monkey kidney cells, and cells were infected with the recombinant vaccinia virus expressing T7 RNA polymerase (vTF7-3) (Fuerst et al. 1986). Other groups used recombinant MVA (modified vaccinia virus Ankara) or a transformed baby hamster kidney BHK cell line to supply T7 RNA polymerase (Leyrer et al. 1998; Nishimura et al. 2007, 2011). After incubation for 3 days in the presence of 1-β-D-arabinofuranosylcytosine (Ara-C) and rifampicin, which were expected to reduce vaccinia virus-associated cytopathogenicity, cell lysates at different dilutions were inoculated into embryonated hen eggs to amplify the recovered recombinant SeV vector. Psoralen compound covalently binds to double-stranded DNA under UV light. UV irradiation of vTF7-3 in the presence of psoralen under certain conditions caused a reduction in vaccinia virus-associated cytopathogenicity, while T7 RNA polymerase expression remained unchanged. As a result, recovery efficiency increased by 100- to 1,000 fold (1 CIU from 10^2 to 10^3 transfected cells) (Li et al. 2000; Hirata et al. 2002, reviewed in Nagai et al. 2007; also see Chap. 2). This technological device greatly facilitated the generation of different kinds of SeV vectors. Final vector titers were comparable to those of recovered SeV without foreign gene insertion, reaching 10^9–10^{10} CIU/mL in egg allantoic fluid and 10^8–10^9 CIU/mL in the supernatant of cells in culture.

3.3.3 Actual Expression Levels of Foreign Genes from the First-Generation SeV Vector

Using the foregoing T7 promoter recovery system, recombinant SeV vectors carrying various foreign genes in the 5′-noncoding region of the *N* gene (*le/N*) (see Sect. 3.3.1) in the full-length SeV genome were successfully recovered and their expression levels in cells in culture were examined. The foreign genes tested for incorporation and expression were those encoding the firefly luciferase, green fluorescent protein (GFP) of *Acquore victoria*, HIV-1 envelope glycoprotein gp120, stromal cell-derived factor 1α (SDF-1α) and SDF-1β of the CXC-chemokine family, β-glucuronidase, β-galactosidase, various cytokines, and neurotrophic factors (Bitzer et al. 2003a, b; Griesenbach et al. 2005; Nagai et al. 2007).

Very high levels of expression were achieved for all these foreign genes. All propagation studies in cell cultures were conducted in the presence of a low concentration of trypsin in the culture media, which was required for the activation of the fusion protein precursor F_0 of the progeny virions (see Sect. 3.2.5). When the luciferase-expressing SeV was inoculated in the CV-1 monkey kidney cell line at an MOI (multiplicity of infection) of 20 (20 CIU/cell), enzyme activity increased exponentially up to 14 h post infection and reached a titer of approximately 3×10^5 cps (count per second) at 20 h. This enzyme activity was about 20 fold higher than that obtained by transfection of the same cells with a luciferase-encoding plasmid driven by T7 RNA polymerase from the recombinant vaccinia virus vTF7-3 (Hasan et al. 1997).

In the case of gp120, expression levels from the standard V(+) SeV reached 2.2 μg/mL in the culture supernatant of infected CV1 cells and V(−) version (see Chap. 2) -based expression was even more marked, consistently reaching more than 6.0 μg/mL, a level that was apparently the highest attainable for gp120 production in mammalian cells (Yu et al. 1997). gp120 was detected as a major protein in the culture supernatant.

In the case of SDF-1α and SDF-1β, expression levels reached as high as 10 μg/mL of the culture supernatant. This level of expression allowed the proteins to be obtained in sufficient quantities to conduct X-ray crystallography (Moriya et al. 1998). The crystal structures, forms 1 and 2, of SDF-1α were determined at 2.0-Å resolution (Ohnishi et al. 2000).

The results of expression of the reporter *Escherichia coli lacZ* gene, which encodes β-galactosidase in the airway epithelial cells of various animal and human origins *in vivo* or ex vivo, were also interesting. Expression levels were shown to be several thousand fold higher than those obtained by the adenovirus vector or cationic liposomes (Yonemitsu et al. 2000). Moreover, the vector expressed a high level of β-galactosidase in *in vivo* and ex vivo tissues such as human vascular endothelial cells (Masaki et al. 2001), skeletal muscle of the rat (Shiotani et al. 2001), joint synovial cells of the rat (Yamashita et al. 2002), retinal tissue of the rat (Ikeda et al. 2002), and the nervous system of gerbils (Shirakura et al. 2003).

The delivery of human clotting Factor IX gene after direct intravenous inoculation of SeV vector in mice was also investigated, and systemic transgene expression of the secreted therapeutic protein was found in the peripheral blood (Bitzer et al. 2003a, b).

3.3.4 *Stability and the Maximum Gene Size Accommodated in the First-Generation SeV Vector*

In general, RNA genomes are more error prone than the DNA genome system. The frequency of nucleotide misincorporation into RNA genomes has been estimated to be as high as 10^{-4} to 10^{-5} per site per replication because of the absence of a proofreading function and repair mechanism in most RNA viruses (Sanjuán et al. 2010). This finding indicates that the mutation rate may be at least several logs higher than that of DNA genomes (10^{-9} to 10^{-10} per site per replication). Therefore, the stability of RNA vectors is a major concern that has been addressed with SeV vectors expressing the firefly luciferase. When the vector was serially passaged in chick embryos by inoculating 10 CIU/egg and amplified to 10^{10} CIU/mL in the allantoic fluid, the expression levels of luciferase did not change until the ninth passage, indicating the stable maintenance of the inserted gene under these conditions (Hasan et al. 1997).

SeV vectors are used after propagation only once or twice in appropriate cells. Thus, the error-prone nature of SeV vectors is practically tolerable. The similar stability of transgene was shown in another nonsegmented negative strand RNA virus (Mebatsion et al. 1996). When inserted into the *le/N* region, SeV was able to accommodate and stably maintain a foreign gene up to 3.2 kb in length (Sakai et al. 1999).

3.4 Second-Generation SeV Vector Created by Deletion of the Viral Essential Genes for Practical Use in Various Settings

Studies on the first-generation SeV vectors in the backbone of the full-length genome suggested that the SeV vector could be a powerful gene transfer tool. However, because vectors can be amplified or mobilized in the body and further

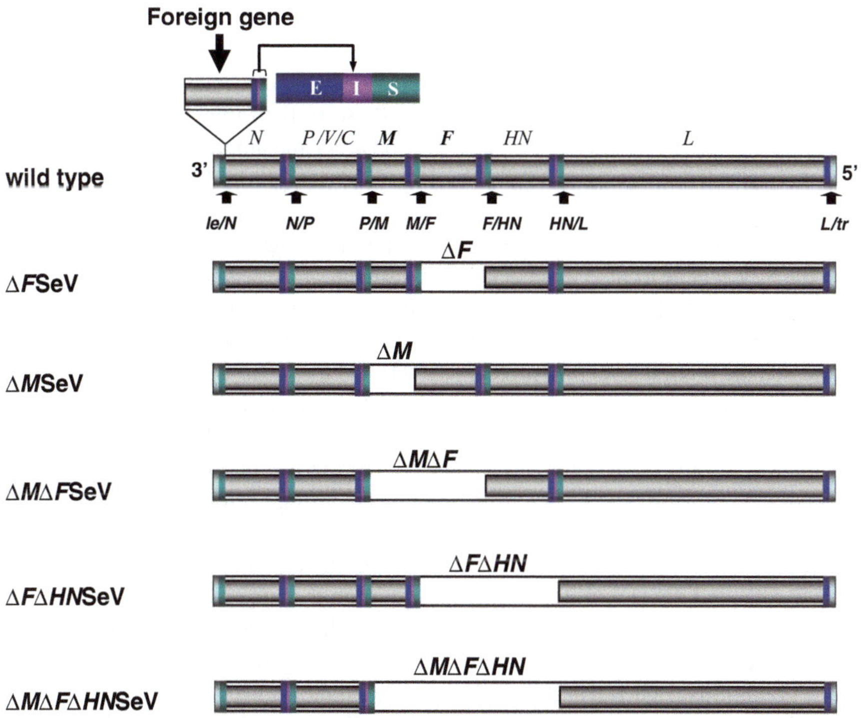

Fig. 3.2 Genome structure of the first-generation SeV vector with a wild-type SeV backbone and various second-generation SeV vectors created by the deletion of the viral essential genes. A foreign gene can be inserted between the start signal and ORF of *N* (*le/N*), *P/V/C* (*N/P*), *M* (*P/M*), *F* (*M/F*), *HN* (*F/HN*), and *L* (*HN/L*) genes, and the termination codon and end signal of the *L* (*L/tr*) gene, respectively. (Adapted from Nagai et al. 2006)

transmitted to other humans, its use in clinical settings for diverse applications necessitated the construction of a nonpropagating, nontransmissible version (Li et al. 2000; reviewed in Griesenbach et al. 2005; Nagai et al. 2007). To fulfill this requirement, we deleted one or more viral essential genes except for the *N*, *P*, and *L* genes, which are indispensable for driving SeV replication (Fig. 3.2). Further engineering was attempted to make the SeV vector compatible for special purposes. One example by us and other groups was deletion of the *P* gene to generate vectors with transient gene expression.

3.4.1 F *Gene-Deleted (Δ*F*) SeV Vector: The First Deletion-Type Vector with No Transmissibility Prepared from the Mononegavirales*

The initial attempt in this line of development was to delete the *F* gene from the SeV genome and insert the EGFP (enhanced green fluorescent protein) gene in the deleted region as a reporter to monitor vector recovery (Li et al. 2000). In addition to the *F* gene-deleted SeV cDNA, the three supporter plasmids encoding the N, P, and L proteins, respectively, were transfected to LLC-MK2 cells to recover RNPs from the *F* gene-deleted SeV cDNA. The recovered RNPs were then transfected into an SeV F-expressing helper cell line (LLC-MK2) to construct and amplify the Δ*F*SeV vector. The helper cell line was established by introducing a plasmid driven by the CAG promoter (a combination of the cytomegalovirus early enhancer element and the chicken β-actin promoter), carrying the G418 resistance gene between two *lox*P sites, and the SeV *F* gene located immediately downstream of the second *lox*P site (Arai et al. 1999). After the isolation of G-418-resistant colonies, Cre recombinase was introduced by a recombinant adenovirus vector carrying the Cre gene into cells, and one clone that expressed the highest level of F protein was selected for use as a helper line.

The *F* gene-deleted (Δ*F*), EGFP-expressing SeV vector isolated and amplified as above had a markedly high production titer (10^7–10^8 CIU/mL). As expected, the vector was confirmed to carry no *F* gene by Northern blot analysis, indicating no reversion to the wild type by acquiring the *F* gene during vector recovery and amplification processes in cells where cDNA of the *F* gene was present. Immunoelectron microscopy clearly revealed the presence of both the F protein originating from the helper cells and the HN protein of a viral origin on the surface of the Δ*F*SeV particles. Although Δ*F*SeV expressed EGFP in initially infected cells, EGFP did not spread to the adjacent cells even in the presence of F-activating trypsin. Under the same conditions, EGFP from the SeV vector with the full-length SeV genome (see Sect. 3.3) spread vigorously in the entire culture. These results indicated that Δ*F*SeV was successfully generated and was nontransmissible.

The transduction efficiency and transgene expression of the Δ*F*SeV vector were markedly high in various types of cells in culture, including primary cells such as human hepatocytes, lung microvascular endothelial cells, and smooth muscle cells (Fig. 3.3), as well as rat cerebral primary neurons and dorsal root ganglia primary neurons (Li et al. 2000). Interestingly, when the vector carrying the EGFP gene was transduced into monkey-derived embryonic stem cells, the cells differentiated normally into embryoid bodies, neurons, and blood cells in a normal manner, indicating that the intracellular presence of the SeV genome does not interfere with cell differentiation (Sasaki et al. 2005). When the vector was transduced *in vivo*, a high level of EGFP or β-galactosidase expression was observed in tissues including the airway epithelium and nasal mucosa epithelial cells of mice, ferrets, and sheep, the skeletal muscle of mice, rats, and monkeys, and the ependymal cells and pyramidal neurons in the rat brain (Fig. 3.4) (Li et al. 2000; Ferrari et al. 2004; Griesenbach et al. 2011). In the mouse fetus, toxicity was lower with the Δ*F*SeV vector than with wild-type SeV (Waddington et al. 2004).

Until now, the Δ*F*SeV vector accommodated, in addition to the EGFP gene, the β-galactosidase gene, luciferase gene, and other genes encoding various enzymes (Griesenbach et al. 2008), cytokines (Inoue et al. 2008), interleukins (Iwadate et al. 2005), neurotrophic factors (Iwasaki et al. 2012), secretory proteins (Tatsuta et al. 2009), membrane proteins (Ban et al. 2007; Suga et al. 2008), and various other intracellular proteins including transcription factors (Kitamura et al. 2011;

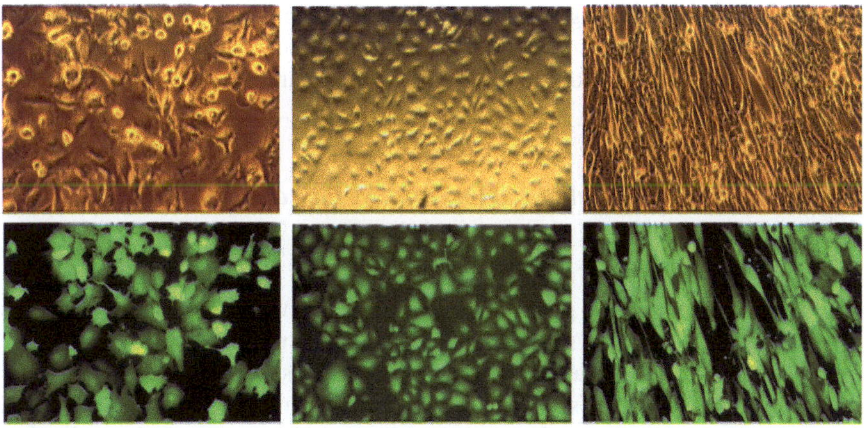

Fig. 3.3 Introduction and expression of the recombinant *F* gene-deleted SeV vector carrying the EGFP gene in various cell types *in vitro*. *Left:* Human hepatocytes. *Middle:* Human lung microvascular endothelial cells. *Right:* Human smooth muscle cells. *Top:* Phase-contrast images. *Bottom:* Fluorescent images of EGFP expression. EGFP expression was observed 3 days after infection with the vector at an MOI of 3. (Adapted from Li et al. 2000)

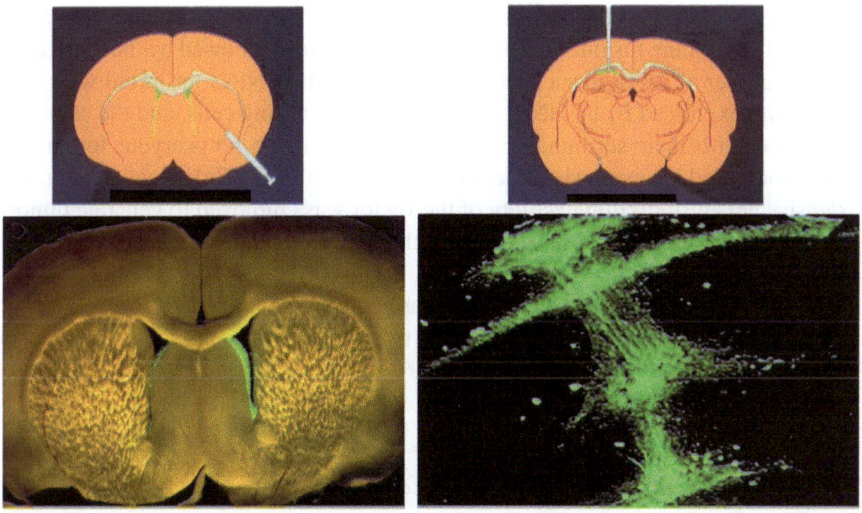

Fig. 3.4 Gene introduction into rat central nervous system. The *F* gene-deleted SeV vector carrying the EGFP gene was stereotaxically injected into rat brain. EGFP expression was observed 4 days after vector injection. *Left:* Gene transfer into the ependymal cells of the lateral ventricle. *Right:* Gene transfer into the pyramidal cells of the CA1 region in the hippocampus. (Adapted from Li et al. 2000)

Fusaki et al. 2009). These genes have been inserted into different positions from *le/N* to *L/tr* to attenuate their expression levels (see Sect. 3.5.2). The maximum capacity of foreign gene accommodation in Δ*F*SeV is, to date, 5.4 kb in size, which is 2.2 kb longer than that of the full-length vector (3.2 kb) (see Sect. 3.3.4, Ban et al. 2007).

For these reasons, including nontransmissibility, high production titers, high levels of transgene expression, and a very broad target cell range, the Δ*F*SeV vector now represents the first in line for routine use for various purposes.

3.4.2 HN *Gene-Deleted* (ΔHN) *and* M *Gene-Deleted* (ΔM) *Versions*

The same protocols used to generate the Δ*F*SeV vector were successfully applied to produce the Δ*HN*SeV and Δ*M*SeV vectors. Their titers ranged from 10^6 to 10^8 CIU/mL. Although Δ*HN*SeV was nontransmissible, which is similar to Δ*F*SeV, Δ*M*SeV was "transmissible", in the strict sense of the word, from cell to cell by a cell-fusion process (Inoue et al. 2003b). Because the M protein is the central player for SeV assembly and particle formation (see Chap. 2; reviewed in Nagai et al. 2011), no progeny virions were formed in Δ*M*SeV infected cells. However, Δ*M*SeV accumulated its RNPs in cells and displayed a large amount of receptor (sialic acid) binding HN and cell fusion inducing F proteins on their surface. When the F proteins were proteolytically activated by adding trypsin to the culture media, the infected cells almost instantly started to fuse with neighboring uninfected cells by F/HN protein-mediated membrane fusion. This phase was immediately followed by the transfer of a large number of RNPs to such neighboring cells. The RNPs transferred were replicated and expressed F and HN proteins in the neighboring cells, and the cells further fused with their surrounding cells, causing massive polykaryote formation, which soon died of apoptosis or necrosis. At the end of Δ*M*SeV development, the extensive cell-fusion/killing activity of the vector was employed to generate tumor-targeting oncolytic vectors for use in virotherapy for malignant solid tumors (see Sect. 3.5.5 and Chap. 5).

Another group also succeeded in generating single gene deletion type SeV vectors using recombinant adenovirus vectors to provide defective SeV gene products in *trans*. The titers of these vectors were 10^6–10^7 CIU/mL (Bernloehr et al. 2004).

3.4.3 *Partial and Complete Deletion of the* P *Gene*

The SeV P protein is an essential cofactor of SeV RNA-dependent RNA polymerase. One of the known roles of the P protein is to recruit soluble N proteins to the nascent RNA chain, which is a critical event to support continued genome replication. The soluble N-binding site of the P protein has been mapped within the N-terminal up to residue 77. Thus, the deletion mutant *P*Δ(2–77) may undergo initial transcription but not subsequent genome replication and amplified transcription (see Chap. 2). A foreign gene was expressed transiently when it was inserted in the SeV vector of this deletion type. This strategy has been employed to generate a vaccine vector against respiratory syncytial virus (see Chap. 4, Bossow et al. 2012).

In our studies, the entire *P* gene-deleted SeV vector (Δ*P*SeV) was generated, which enabled very rapid clearance of the vector as well as the transgene from the vector-transduced cells. When the homeobox B4 (HoxB4) gene, which promotes the expansion of hematopoietic stem cells (HSCs), was introduced into human cord blood $CD34^+$ cells using the Δ*P*SeV vector, $CD34^+$ cells expanded well ex vivo. These expanded cells were then transplanted into the abdominal cavity of fetal sheep. Interestingly, HoxB4-transduced $CD34^+$HSCs were shown to contribute to repopulation in sheep up to 20 months (Abe et al. 2011).

3.4.4 *Double and Triple Deletion*

Further created were doubly deleted (Δ*M*Δ*F* and Δ*F*ΔHN) and triply deleted (Δ*M*Δ*F*Δ*HN*) SeV vectors (Inoue et al. 2004; Yoshizaki et al. 2006; Tanaka et al. 2007; Murakami et al. 2008; reviewed in Nagai et al. 2007). When amplified in the respective helper cells, their production titers were 10^6–10^7 CIU/mL. No production of transmissible particles was observed in normal cells for these deletion mutants. The triple deletion vector was attenuated in cytopathogenicity *in vitro* and *in vivo* (Yoshizaki et al. 2006) and showed reduced innate immunogenicity in mice (Murakami et al. 2008).

3.5 Some Essential Improvements in SeV Technology for Various Applications

3.5.1 *Vaccinia Virus-Free System for Vector Recovery*

We attempted to establish a vaccinia virus-free system to avoid the risk of contamination with the helper vaccinia virus in the resulting SeV vectors. All the supporting plasmids to express the N, P, and L proteins and an additional plasmid to express T7 RNA polymerase were driven by cellular RNA polymerase II under the control of the CAG promoter. As a result of various trials to maintain high efficiency vector recovery, the plasmid to generate SeV genome RNA was left unchanged and was under the control of the T7 promoter. The resulting all-plasmid-driven method demonstrated virus recovery efficiency similar to or even higher than that obtained by the UV-psoralen vTF7-3 system (see Sect. 3.3.2). When a gene deletion-type SeV vector was required, another plasmid carrying the deleted gene under the control of the T7 or CAG promoter was added to the system (see Sect. 3.4).

3.5.2 Regulation of the Level of Foreign Gene Expression by Selecting the Insertion Site in the SeV Genome

A strong requirement exists to control gene expression levels because (1) the expressed proteins must be at a high enough or appropriate concentration to exert their functions and (2) the overexpression of some products may cause toxicity in the transduced cells. Gene expression by retroviral, adenoviral, and adeno-associated viral vectors can be controlled by *cis*-elements on the DNA stretch such as the promoter/enhancer system recognized by cellular DNA-dependent RNA polymerase. In contrast, no such controlling mechanisms exist for SeV gene expression. Instead, two virus-specific mechanisms are plausible for the control of SeV gene expression. One is by the different reinitiation capacity of transcription because of subtle nucleotide changes in the start signals. Specifically, the reinitiation capacity of the *F*-gene start was shown to be about one fifth of those of the other start signals (Kato et al. 1999; reviewed in Nagai et al. 2011; also see Chap. 2). The other is by the so-called "polar attenuation". Even though the reinitiation signal is strong enough, the entrance of the polymerase to the next start signal may not be perfect. Therefore, gradual transcription attenuation is believed to occur toward the 5′-end (Homann et al. 1990). In this model, the expression level of the *N* gene, which is most proximal to the 3′-end, is the highest, whereas that of the most 5′ proximal *L* gene is the lowest (see Chap. 2).

We investigated how far these plausible mechanisms were applicable to the control of foreign gene expression. A series of SeV vectors were generated that carried a *Not*I insertion site in the noncoding regions between the start signal and the initiation codon of the *N* (see Sect. 3.3.1), *P*, *M*, *F*, *HN*, and *L* genes in addition to the site between the termination codon and the stop signal in the *L* gene followed by the 5′-trailer (*tr*) sequence. These insertion sites have been defined as *le*/*N* (see Sect. 3.3.1), *N/P*, *P/M*, *M/F*, *F/HN*, *HN/L*, and *L/tr*, respectively. The synthetic E-I-S was placed before the *Not*I tag in the antisense primer for the former six constructs and after the *Not*I tag in the sense primer for *L/tr*. All vectors were successfully generated, and into these a reporter gene encoding SEAP (secreted embryonic alkaline phosphatase) was inserted. A polar, stepwise attenuation of the reporter gene was found toward the 5′-end of the SeV genome in infected CV1 cells at an MOI of 20 (20 CIU/cell) (Tokusumi et al. 2002). Expression from *le*/*N* was about 20 times higher than that from *L*/tr. This attenuation was not linear, but was biphasic because of the weaker reinitiation capacity of the *F*-gene start, resulting in a greater attenuation of the *F/HN* and downstream inserts (reviewed in Nagai et al. 2011; also see Chap. 2).

These results indicated that foreign gene expression from the SeV vector could be up- or downregulated by selecting an appropriate insertion site. This control method has an advantage in that mechanisms are operated by SeV itself and are independent of the physiological state of transduced cells.

3.5.3 Introduction of Temperature-Sensitive Mutations to Generate "Footprint-Free" SeV Vectors

To date, various temperature-sensitive (ts) mutants of SeV have been isolated and the responsible ts mutations in the SeV genome have been identified (Adachi et al. 1980; Bowman et al. 1999; Feller et al. 2000). Among them, point mutations in the *P* and *L* genes conferred clear-cut ts phenotypes. For example, one such mutant maintained gene expression and genome replication at 35 °C, but not at 37 °C, whereas another was active at 37 °C, but not at 39 °C. The introduction of various combinations of such mutations into the Δ*F*SeV vector allowed the generation of "footprint-free" vectors, which underwent foreign gene expression at a permissive temperature and then were erased from the cells by shifting up to a nonpermissive temperature (Ban et al. 2011). This system has been successfully applied to generate induced pluripotent stem cells (iPSCs) free of reprogramming factors (see Chap. 7).

3.5.4 Generation of Vectors with Reduced Cytotoxicity

A Δ*F*SeV vector causing reduced cytotoxicity was developed by introducing additional ts mutations found in the *P*, *M*, *HN*, and *L* genes (Inoue et al. 2003a, b; Ban et al. 2011). A vector of this type exhibited longer-term gene expression in various transduced cells *in vitro*. The use of this vector for cell reprogramming has been discussed in Chap. 7. A previous report also described the preparation of recombinant SeV with reduced cytotoxicity by replacing the first 42 nt of the leader sequence with the corresponding sequence of the trailer region. This recombinant SeV did not cause programmed cell death and produced a stable persistent infection (Iseni et al. 2002; Wiegand et al. 2005). More recently, the construction of vectors based on SeV isolated from persistently infected cells has been reported (Nishimura et al. 2007).

3.5.5 Changing Target Cells Based on the Protease-Dependent Theory of SeV Tropism

In nature, SeV targets limited types of tissues in the host organism in spite of its use as a receptor for the terminal sialic acids ubiquitously present in the animal body. This phenomenon is well attributable to the fact that the inactive precursor F_0 protein of SeV is cleaved into the biologically active F_1 and F_2 subunits only by an endoprotease highly specific to certain tissues, and not by a ubiquitous endoprotease. The virus is thought to target only the respiratory tracts of mice, and does not spread to other tissues and organs because the responsible protease(s) is expressed on the airway surface, but not in other tissues (reviewed in Nagai 1993). In the chick embryo, SeV spreads only in the cells surrounding the allantoic or amniotic cavities and cannot spread beyond these sites. In this egg model, the blood clotting factor FXa was identified as the F_0 protein-activating protease (Gotoh et al. 1990; Suzuki et al. 1991). FXa is secreted from the allantoic or amniotic cells and accumulates in the embryonic cavities but was absent in other embryonic tissues (Ogasawara et al. 1992; reviewed in Nagai 1993; also see Chap. 2). In cell cultures, SeV spreads by multiple cycles of replication only in the presence of a low concentration of trypsin, which can be substituted for a naturally occurring protease.

The conversion of the SeV F_0 cleavage site (Q–S–R) to a site cleavable by proteases in other category would render SeV F_0 cleavable by that category of proteases and may undergo multiple cycles of replication where that category of proteases is expressed. Such manipulation was successfully performed so that the F_0 protein of ΔMSeV underwent cleavage activation by matrix metalloproteases (MMPs) or a urokinase-type plasminogen activator (uPA), both of which are overexpressed by certain malignant tumor cells (Kinoh et al. 2004, 2009; Hasegawa et al. 2010; Morodomi et al. 2012). This method has provided a new approach to oncolytic virotherapy (see Chap. 6).

3.6 Conclusions

The SeV vector is a novel class of expression vector whose concepts are different in various important aspects from those of other viral vectors previously developed. The vector is a multipurpose cytoplasmic, nonintegrating RNA replicon vector with a markedly high capacity of transgene expression and a very broad target cell range. Engineering of the SeV genome has advanced, with the creation of safer versions for general purposes and special versions for specific purposes. Further devices will enable SeV technology to be applied for more diverse medical needs. These devices may include ways to reduce or minimize immunogenicity and achieve more selective targeting.

Acknowledgments We are grateful to M. Hasegawa for critical reading of the manuscript. We thank H. Hara for his careful reading of the manuscript and assistance in English usage.

References

Abe T, Masuda S, Ban H, Hayashi S, Ueda Y, Inoue M, Hasegawa M, Nagao Y, Hanazono Y (2011) Ex vivo expansion of human HSCs with Sendai virus vector expressing HoxB4 assessed by sheep in utero transplantation. Exp Hematol 39(1):47–54

Adachi A, Kanda T, Shibuta H (1980) Isolation and characterization of temperature-sensitive mutants of Sendai virus. Microbiol Immunol 24(11):1053–1068

Arai T, Matsumoto K, Saitoh K, Ui M, Ito T, Murakami M, Kanegae Y, Saito I, Cosset FL, Takeuchi Y, Iba H (1999) A new system for stringent, high-titer vesicular stomatitis virus G protein-pseudotyped retrovirus vector induction by introduction of Cre recombinase into stable prepackaging cell lines. J Virol 72(2):1115–1121

Armeanu S, Ungerechts G, Bernloehr C, Bossow S, Gregor M, Neubert WJ, Lauer UM, Bitzer M (2003) Cell cycle independent infection and gene transfer by recombinant Sendai viruses. J Virol Methods 108:229–233

Ban H, Inoue M, Griesenbach U, Munkonge F, Chan M, Iida A, Alton EW, Hasegawa M (2007) Expression and maturation of Sendai virus vector-derived CFTR protein: functional and biochemical evidence using a GFP-CFTR fusion protein. Gene Ther 14(24):1688–1694

Ban H, Nishishita N, Fusaki N, Tabata T, Saeki K, Shikamura M, Takada N, Inoue M, Hasegawa M, Kawamata S, Nishikawa S (2011) Efficient generation of transgene-free human induced pluripotent stem cells (iPSCs) by temperature-sensitive Sendai virus vectors. Proc Natl Acad Sci USA 108(34):14234–14239

Bernloehr C, Bossow S, Ungerechts G, Armeanu S, Neubert WJ, Lauer UM, Bitzer M (2004) Efficient propagation of single gene deleted recombinant Sendai virus vectors. Virus Res 99(2):193–197

Bitzer M, Armeanu S, Lauer UM, Neubert WJ (2003a) Sendai virus vectors as an emerging negative-strand RNA viral vector system. J Gene Med 5:543–553

Bitzer M, Ungrechts G, Bossow S, Graepler F, Sedlmeier R, Ameanu S, Bernloehr C, Spiegel M (2003b) Negative-strand RNA viral vectors: intravenous application of Sendai virus vectors for the systemic delivery of therapeutic genes. Mol Ther 7:210–217

Bossow S, Schlecht S, Schubbert R, Pfeiffer M, Neubert WJ, Wiegand M (2012) Evaluation of nucleocapsid and phosphoprotein P functionality as critical factors during the early phase of paramyxoviral infection. Open Virol J 6:73–81

Bowman MC, Smallwood S, Moyer SA (1999) Dissection of individual functions of the Sendai virus phosphoprotein in transcription. J Virol 73(8):6474–6483

Calain P, Roux L (1993) The rule of six, a basic feature for efficient replication of Sendai virus defective interfering RNA. J Virol 67:4822–4830

Chare ER, Gould EA, Holmes EC (2003) Phylogenetic analysis reveals a low rate of homologous recombination in negative-sense RNA viruses. J Gen Virol 84(Pt 10):2691–2703

Conzelmann K-K (1998) Nonsegmented negative-strand RNA viruses. Annu Rev Genet 32:123–162

Feller JA, Smallwood S, Skiadopoulos MH, Murphy BR, Moyer SA (2000) Comparison of identical temperature-sensitive mutations in the L polymerase proteins of Sendai and parainfluenza 3 viruses. Virology 276(1):190–201

Ferrari S, Griesenbach U, Shiraki-Iida T, Shu T, Hironaka T, Hou X, Williams J, Zhu J, Jeffery PK, Geddes DM, Hasegawa M, Alton EW (2004) A defective nontransmissible recombinant Sendai virus mediates efficient gene transfer to airway epithelium in vivo. Gene Ther 11(22): 1659–1664

Fuerst TR, Niles FW, Studier FW, Moss B (1986) Eukaryotic transient-expression system based on recombinant vaccinia virus that synthesizes bacteriophage T7 RNA polymerase. Proc Natl Acad Sci USA 83:8122–8126

Fusaki N, Ban H, Nishiyama A, Saeki K, Hasegawa M (2009) Efficient induction of transgene-free human pluripotent stem cells using a vector based on Sendai virus, an RNA virus that does not integrate into the host genome. Proc Jpn Acad Ser B Phys Biol Sci 85(8):348–362

Gotoh B, Ogasawara T, Toyoda T, Inocencio NM, Hamaguchi M, Nagai Y (1990) An endoprotease homologous to the blood clotting factor X as a determinant of viral tropism in chick embryo. EMBO J 9(12):4189–4195

Griesenbach U, Cassady RL, Ferrari S, Fukumura M, Müller C, Schmitt E, Zhu J, Jeffery PK, Nagai Y, Geddes DM, Hasegawa M, Alton EW (2002) The nasal epithelium as a factory for systemic protein delivery. Mol Ther 5(2):98–103

Griesenbach U, Inoue M, Hasegawa M, Alton EWFW (2005) Sendai virus for gene therapy and vaccination. Curr Opin Mol Ther 7:346–352

Griesenbach U, Meng C, Farley R, Cheng SH, Scheule RK, Davies MH, Wolstenholme-Hogg PC, ten Hove W, van der Hoeven P, Sinn PL, McCray PB Jr, Inoue M, Geddes DM, Hasegawa M, Frankel G, Wiles S, Alton EW (2008) In vivo imaging of gene transfer to the respiratory tract. Biomaterials 29(10):1533–1540

Griesenbach U, McLachlan G, Owaki T, Somerton L, Shu T, Baker A, Tennant P, Gordon C, Vrettou C, Baker E, Collie DD, Hasegawa M, Alton EW (2011) Validation of recombinant Sendai virus in a non-natural host model. Gene Ther 18(2):182–188

Hacein-Bey-Abina S, Von Kalle C, Schmidt M, McCormack MP, Wulffraat N, Leboulch P, Lim A, Osborne CS, Pawliuk R, Morillon E, Sorensen R, Forster A, Fraser P, Cohen JI, de Saint BG, Alexander I, Wintergerst U, Frebourg T, Aurias A, Stoppa-Lyonnet D, Romana S, Radford-Weiss I, Gross F, Valensi F, Delabesse E, Macintyre E, Sigaux F, Soulier J, Leiva LE, Wissler M, Prinz C, Rabbitts TH, Le Deist F, Fischer A, Cavazzana-Calvo M (2003) LMO2-associated clonal T cell proliferation in two patients after gene therapy for SCID-X1. Science 302(5644):415–419

Hacein-Bey-Abina S, Garrigue A, Wang GP, Soulier J, Lim A, Morillon E, Clappier E, Caccavelli L, Delabesse E, Beldjord K, Asnafi V, MacIntyre E, Dal Cortivo L, Radford I, Brousse N, Sigaux F, Moshous D, Hauer J, Borkhardt A, Belohradsky BH, Wintergerst U, Velez MC, Leiva L, Sorensen R, Wulffraat N, Blanche S, Bushman FD, Fischer A, Cavazzana-Calvo M (2008) Insertional oncogenesis in 4 patients after retrovirus-mediated gene therapy of SCID-X1. J Clin Invest 118(9):3132–3142

Hara H, Hara H, Hironaka T, Inoue M, Iida A, Shu T, Hasegawa M, Nagai Y, Falsey AR, Kamali A, Anzala O, Sanders EJ, Karita E, Mwananyanda L, Vasan S, Lombardo A, Parks CL, Sayeed E, Krebs M, Cormier E, Ackland J, Price MA, Excler JL (2011) Prevalence of specific neutralizing antibodies against Sendai virus in populations from different geographic areas: implications for AIDS vaccine development using Sendai virus vectors. Hum Vaccin 7(6):639–645

Hasan MK, Kato A, Shioda T, Sakai Y, Yu D, Nagai Y (1997) Creation of an infectious recombinant Sendai virus expressing the firefly luciferase from 3′ proximal first locus. J Gen Virol 78:2813–2820

Hasegawa Y, Konoh H, Iwadate Y, Onimaru M, Ueda Y, Harada Y, Saito S, Furuya A, Saegusa T, Morodomi Y, Hasegawa M, Saito S, Aoki I, Saeki N, Yonemitsu Y (2010) Urokinase-targeted fusion by oncolytic Sendai virus eradicates orthotopic glioblastomas by pronounced synergy with interferon-β gene. Mol Ther 18:1778–1786

Hirata T, Iida A, Shiraki-Iida T, Kitazato K, Kato A, Nagai Y, Hasegawa M (2002) The improved method for recovery of F-defective Sendai virus expressing foreign genes from cloned cDNA. J Virol Methods 104(2):125–133

Homann HE, Hofschneider PH, Neubert WJ (1990) Sendai virus gene expression in lytically and persistently infected cells. Virology 177:131–140

Homma M, Ohuchi M (1973) Trypsin action on the growth of Sendai virus in tissue culture cells III. Structural difference of Sendai viruses grown in eggs and tissue culture cells. J Virol 12(6):1457–1465

Ikeda Y, Yonemitsu Y, Sakamoto T, Ishibashi T, Ueno H, Kato A, Nagai Y, Fukumura M, Inomata H, Hasegawa M, Sueishi K (2002) Recombinant Sendai virus-mediated gene transfer into adult rat retinal tissue: efficient gene transfer by brief exposure. Exp Eye Res 75(1):39–48

Inoue M, Tokusumi Y, Ban H, Kanaya T, Tokusumi T, Nagai Y, Iida A, Hasegawa M (2003a) Non-transmissible virus-like particle formation by F-deficient Sendai virus is temperature-sensitive and reduced by mutations in M and HN proteins. J Virol 77(5):3238–3246

Inoue M, Tokusumi Y, Ban H, Kanaya T, Shirakura M, Tokusumi T, Hirata T, Nagai Y, Iida A, Hasegawa M (2003b) A new type of Sendai virus vector deficient in the matrix gene has lost virus particle formation and gained extensive cell to cell spreading. J Virol 77(11):6419–6429

Inoue M, Tokusumi Y, Ban H, Shirakura M, Kanaya T, Yoshizaki M, Hironaka T, Nagai Y, Iida A, Hasegawa M (2004) Recombinant Sendai virus vectors deleted in both the matrix and the fusion genes: efficient gene transfer with preferable properties. J Gene Med 6(10):1069–1081

Inoue H, Iga M, Nabeta H, Yokoo T, Suehiro Y, Okano S, Inoue M, Kinoh H, Katagiri T, Takayama K, Yonemitsu Y, Hasegawa M, Nakamura Y, Nakanishi Y, Tani K (2008) Non-transmissible Sendai virus encoding granulocyte macrophage colony-stimulating factor is a novel and potent vector system for producing autologous tumor vaccines. Cancer Sci 99(11):2315–2326

Iseni F, Garcin D, Nishio M, Kedersha N, Anderson P, Kolakofsky D (2002) Sendai virus trailer RNA binds TIAR, a cellular protein involved in virus-induced apoptosis. EMBO J 21(19):5141–5150

Iwadate Y, Inoue M, Saegusa T, Tokusumi Y, Kinoh H, Hasegawa M, Tagawa M, Yamaura A, Shimada H (2005) Recombinant Sendai virus vector induces complete remission of established brain tumors through efficient interleukin-2 gene transfer in vaccinated rats. Clin Cancer Res 11(10):3821–3827

Iwasaki Y, Negishi T, Inoue M, Tashiro T, Tabira T, Kimura N (2012) Sendai virus vector-mediated brain-derived neurotrophic factor expression ameliorates memory deficits and synaptic degeneration in a transgenic mouse model of Alzheimer's disease. J Neurosci Res 90(5):981–989

Kato A, Sakai Y, Shioda T, Kondo T, Nakanishi M, Nagai Y (1996) Initiation of Sendai virus multiplication from transfected cDNA or RNA with negative or positive sense. Genes Cells 1:569–579

Kato A, Kiyotani K, Hasan MK, Shioda T, Sakai Y, Yoshida T, Nagai Y (1999) Sendai virus gene start signals are not equivalent in reinitiation capacity: moderation at the fusion protein gene. J Virol 73(11):9237–9246

Kinoh H, Inoue M, Washizawa K, Yamamoto T, Fujikawa S, Tokusumi Y, Iida A, Nagai Y, Hasegawa M (2004) Generation of a recombinant Sendai virus that selectively activated and lyse human tumor cells expressing matrix metalloproteinases. Gene Ther 11:1137–1145

Kinoh H, Inoue M, Komaru A, Ueda Y, Hasegawa M, Yonemitsu Y (2009) Generation of optimized and urokinase-targeted oncolytic Sendai virus vectors applicable for various human malignancies. Gene Ther 16:392–403

Kitamura A, Matsushita K, Takiguchi Y, Shimada H, Tada Y, Yamanaka M, Hiroshima K, Tagawa M, Tomonaga T, Matsubara H, Inoue M, Hasegawa M, Sato Y, Levens D, Tatsumi K, Nomura F (2011) Synergistic effect of non-transmissible Sendai virus vector encoding the c-*myc* suppressor FUSE-binding protein-interacting repressor plus cisplatin in the treatment of malignant pleural mesothelioma. Cancer Sci 102(7):1366–1373

Kolakofsky D, Pelet T, Garcin D, Hausmann S, Curran J, Roux L (1998) Paramyxovirus RNA synthesis and the requirement for hexamer genome length: the rule of six revisited. J Virol 72:891–899

Komada H, Kusagawa S, Orvell C, Tsurudome M, Nishio M, Bando H, Kawano M, Matsumura H, Norrby E, Ito Y (1992) Antigenic diversity of human parainfluenza virus type 1 isolates and their immunological relationship with Sendai virus revealed by using monoclonal antibodies. J Gen Virol 73(Pt 4):875–884

Kurihara K, Takahara Y, Nomura T, Ishii H, Iwamoto N, Takahashi N, Inoue M, Iida A, Hara H, Shu T, Hasegawa M, Moriya C, Matano T (2012) Immunogenicity of repeated Sendai viral vector vaccination in macaques. Microbes Infect 14(13):1169–1176

Lamb RA, Parks GD (2007) Paramyxoviridae: the viruses and their replication. In: Knipe DM, Howley PM (eds) Fields virology, vol 1, 5th edn. Lippincott-Raven, Philadelphia

Leyrer S, Neubert WJ, Sedlmeier R (1998) Rapid and efficient recovery of Sendai virus from cDNA: factors influencing recombinant virus rescue. J Virol Methods 75(1):47–58

Li H, Zhu Y, Asakawa M, Kuma H, Hirata T, Ueda Y, Lee Y, Fukumura M, Iida A, Kato A, Nagai Y, Hasegawa M (2000) A cytoplasmic RNA vector derived from nontransmissible Sendai virus with efficient gene transfer and expression. J Virol 74:6564–6569

Masaki I, Yonemitsu Y, Komori K, Ueno H, Nakashima Y, Nakagawa K, Fukumura M, Kato A, Hasan MK, Nagai Y, Sugimachi K, Hasegawa M, Sueishi K (2001) Recombinant Sendai virus-mediated gene transfer to vasculature: a new class of efficient gene transfer vector to the vascular system. FASEB J 15:1294–1296

Mebatsion T, Schnell MJ, Cox JH, Finke S, Conzelmann KK (1996) Highly stable expression of a foreign gene from rabies virus vectors. Proc Natl Acad Sci USA 93(14):7310–7314

Moriya C, Shioda T, Tashiro K, Nagasawa T, Ikegawa M, Ohnishi Y, Kato A, Hu H, Xin X, Hasan MK, Maekawa M, Takebe Y, Sakai Y, Honjo T, Nagai Y (1998) Large quantity production with extreme convenience of human SDF-1α and SDF-1β by a Sendai virus vector. FEBS Lett 425(1):105–111

Morodomi Y, Yano T, Kinoh H, Harada Y, Saito S, Kyuragi R, Yoshida K, Onimaru M, Shoji F, Yoshida T, Ito K, Shikada Y, Maruyama R, Hasegawa M, Maehara Y, Yonemitsu Y (2012)

BioKnife, a uPA activity-dependent oncolytic Sendai virus, eliminates pleural spread of malignant mesothelioma via simultaneous stimulation of uPA expression. Mol Ther 20(4):769–777
Murakami Y, Ikeda Y, Yonemitsu Y, Tanaka S, Kondo H, Okano S, Kohno R, Miyazaki M, Inoue M, Hasegawa M, Ishibashi T, Sueishi K (2008) Newly-developed Sendai virus vector for retinal gene transfer: reduction of innate immune response via deletion of all envelope-related genes. J Gene Med 10(2):165–176
Nagai Y (1993) Protease-dependent virus tropism and pathogenicity. Trends Microbiol 1(3): 81–87
Nagai Y, Kato A (1999) Paramyxovirus reverse genetics is coming of age. Microbiol Immunol 43:613–624
Nagai Y, Kato A, Inoue M (2006) Establishment of progress of Sendai virus engineering. Tanpakushitu Kakusan Koso 51(1):27–37 (Review, Japanese)
Nagai Y, Inoue M, Iida A, Zhu Y-F, Hasegawa M, Kato A, Matano T (2007) Sendai virus engineering: from reverse genetics to vector development. In: Hefferon KL (ed) Virus expression vectors. Transworld Research Network, Kerala
Nagai Y, Takakura A, Irie T, Yonemitsu M, Gotoh B (2011) Evolution from mouse pathogen to a state-of-the-art tool in virus research and biotechnology. In: Samal SK (ed) The paramyxoviruses. Horizon Scientific Press, Norwich
Nishimura K, Segawa H, Goto T, Morishita M, Masago A, Takahashi H, Ohmiya Y, Sakaguchi T, Asada M, Imamura T, Shimotono K, Takayama K, Yoshida T, Nakanishi M (2007) Persistent and stable gene expression by a cytoplasmic RNA replicon based on a noncytopathic variant Sendai virus. J Biol Chem 282:27383–27391
Nishimura K, Sano M, Ohtake M, Furuta B, Umemura Y, Nakajima Y, Ikehara Y, Kobayashi T, Segawa H, Takagasu S, Sato H, Motomura K, Uchida E, Kanayasu-Toyoda T, Asashima M, Nakauchi H, Yamaguchi T, Nakanishi M (2011) Development of defective and persistent Sendai virus vector: a unique gene delivery/expression system ideal for cell reprogramming. J Biol Chem 286(6):4760–4771
Ogasawara T, Gotoh B, Suzuki H, Asaka J, Shimokata K, Rott R, Nagai Y (1992) Expression of factor X and its significance for the determination of paramyxovirus tropism in the chick embryo. EMBO J 11:467–472
Ohnishi Y, Senda T, Nandhagopal N, Sugimoto K, Shioda T, Nagal Y, Mitsui Y (2000) Crystal structure of recombinant native SDF-1α with additional mutagenesis studies: An attempt at a more comprehensive interpretation of accumulated structure-activity relationship data. J Interferon Cytokine Res 20(8):691–700
Re GG, Morgan EM, Kingsbury DW (1985) Nucleotide sequences responsible for generation of internally deleted Sendai virus defective interfering genomes. Virology 146(1):27–37
Sakai Y, Kiyotani K, Fukumura M, Asakawa M, Kato A, Shioda T, Yoshida T, Tanaka A, Hasegawa M, Nagai Y (1999) Accommodation of foreign genes into the Sendai virus genome: sizes of inserted genes and viral replication. FEBS Lett 456(2):221–226
Sanjuán R, Nebot MR, Chirico N, Mansky LM, Belshaw R (2010) Viral mutation rates. J Virol 84(19):9733–9748
Sasaki K, Inoue M, Shibata H, Ueda Y, Muramatsu SI, Okada T, Hasegawa M, Ozawa K, Hanazono Y (2005) Efficient and stable Sendai virus-mediated gene transfer into primate embryonic stem cells with pluripotency preserved. Gene Ther 12:203–210
Shiotani A, Fukumura M, Maeda M, Hou X, Inoue M, Kanamori T, Komaba S, Washizawa K, Fujikawa S, Yamamoto T, Kadono C, Watabe K, Fukuda H, Saito K, Sakai Y, Nagai Y, Kanzaki J, Hasegawa M (2001) Skeletal muscle regeneration after insulin-like growth factor I gene transfer by recombinant Sendai virus vector. Gene Ther 8(14):1043–1050
Shirakura M, Fukumura M, Inoue M, Fujikawa S, Maeda M, Watabe K, Kyuwa S, Yoshikawa Y, Hasegawa M (2003) Sendai virus vector-mediated gene transfer of glial cell line-derived neurotrophic factor prevents delayed neuronal death after transient global ischemia in gerbils. Exp Anim 52(2):119–127
Simon-Loriere E, Holmes EC (2011) Why do RNA viruses recombine? Nat Rev Microbiol 9:617–626

Slobod KS, Shenep JL, Luján-Zilbermann J, Allison K, Brown B, Scroggs RA, Portner A, Coleclough C, Hurwitz JL (2004) Safety and immunogenicity of intranasal murine parainfluenza virus type 1 (Sendai virus) in healthy human adults. Vaccine 22:3182–3186

Suga H, Nagasaki H, Kondo TA, Okajima Y, Suzuki C, Ozaki N, Arima H, Yamamoto T, Ozaki N, Akai M, Sato A, Uozumi N, Inoue M, Hasegawa M, Oiso Y (2008) Novel treatment for lithium-induced nephrogenic diabetes insipidus rat model using the Sendai-virus vector carrying aquaporin 2 gene. Endocrinology 149(11):5803–5810

Suzuki H, Harada A, Hayashi Y, Wada K, Asaka J, Gotoh B, Ogasawara T, Nagai Y (1991) Primary structure of the virus activating protease from chick embryo. Its identity with the blood clotting factor Xa. FEBS Lett 283(2):281–285

Tanaka S, Yonemitsu Y, Yoshida K, Okano S, Kondo H, Inoue M, Hasegawa M, Masumoto K, Suita S, Taguchi T, Sueishi K (2007) Impact of deletion of envelope-related genes of recombinant Sendai viruses on immune responses following pulmonary gene transfer of neonatal mice. Gene Ther 14(13):1017–1028

Tapparel C, Maurice D, Roux L (1998) The activity of Sendai virus genomic and antigenomic promoters requires a second element past the leader template regions: a motif, $(GNNNNN)_3$, is essential for replication. J Virol 72:3117–3128

Tatsuta K, Tanaka S, Tajiri T, Shibata S, Komaru A, Ueda Y, Inoue M, Hasegawa M, Suita S, Sueishi K, Taguchi T, Yonemitsu Y (2009) Complete elimination of established neuroblastoma by synergistic action of gamma-irradiation and DCs treated with rSeV expressing interferon-beta gene. Gene Ther 16(2):240–251

Thomas CE, Ehrhardt A, Kay MA (2003) Progress and problems with the use of viral vectors for gene therapy. Nat Rev Genet 4(5):346–358

Tokusumi T, Iida A, Hirata T, Kato A, Nagai Y, Hasegawa M (2002) Recombinant Sendai viruses expressing different levels of a foreign reporter gene. Virus Res 86:33–38

Verma I, Weitzman M (2005) Gene therapy: twenty-first century medicine. Annu Rev Biochem 74:711–738

Vulliemoz D, Roux L (2001) "Rule of six:" how does the Sendai virus RNA polymerase keep count? J Virol 75:4506–4518

Waddington SN, Buckley SM, Bernloehr C, Bossow S, Ungerechts G, Cook T, Gregory L, Rahim A, Themis M, Neubert WJ, Coutelle C, Lauer UM, Bitzer M (2004) Reduced toxicity of F-deficient Sendai virus vector in the mouse fetus. Gene Ther 11(7):599–608

Walters RW, Grunst T, Bergelson JM, Finberg RW, Welsh MJ, Zabner J (1999) Basolateral localization of fiber receptors limits adenovirus infection from the apical surface of airway epithelia. J Biol Chem 274(15):10219–10226

Wiegand M, Bossow S, Neubert WJ (2005) Sendai virus trailer RNA simultaneously blocks two apoptosis-inducing mechanisms in a cell type-dependent manner. J Gen Virol 86(pt 8): 2305–2314

Yamashita A, Yonemitsu Y, Okano S, Nakagawa K, Nakashima Y, Irisa T, Iwamoto Y, Nagai Y, Hasegawa M, Sueishi K (2002) Fibroblast growth factor-2 determines severity of joint disease in adjuvant-induced arthritis in rats. J Immunol 168(1):450–457

Yonemitsu Y, Kitson C, Ferrari S, Farley R, Griesenbach U, Judd DJ, Steel R, Scheid P, Zhu J, Jeffery PK, Kato A, Hasan MK, Masaki I, Nagai Y, Fukumura M, Hasegawa M, Geddes DM, Alton EWFW (2000) Efficient gene transfer to the airway epithelium using recombinant Sendai virus. Nat Biotechnol 18:970–973

Yonemitsu Y, Matsumoto T, Itoh H, Okazaki J, Uchiyama M, Yoshida K, Onimaru M, Onohara T, Inoguchi H, Kyuragi R, Guntani A, Shimokawa M, Ban H, Tanaka M, Inoue M, Shu T, Hasegawa M, Nakanishi Y, Maehara Y (2013) DVC1-0101 to treat peripheral arterial disease: phase I/IIa, open-label, dose escalation clinical trial. Mol Ther 21(3):707–714

Yoshizaki M, Hironaka T, Iwasaki H, Ban H, Tokusumi Y, Iida A, Nagai Y, Hasegawa M, Inoue M (2006) Naked Sendai virus vector lacking all of the envelope-related gene: reduced cytopathogenicity and immunogenicity. J Gene Med 8:1151–1159

Yu D, Shioda T, Kato A, Hasan MK, Sakai Y, Nagai Y (1997) Sendai virus-based expression of HIV-1 gp120: reinforcement by the V(−) version. Genes Cells 2:457–466

Chapter 4
Genome Replication-Incompetent Sendai Virus Vaccine Vector Against Respiratory Viral Infections That Is Capable of Eliciting a Broad Spectrum of Specific Immune Response

Marian Wiegand and Wolfgang J. Neubert

Abstract Vaccines have proven to be the most effective measure against infectious diseases. However, not every infectious disease can be prevented by vaccination at the present time. For some diseases, new vaccination strategies have to be developed because classical approaches have not been successful. Viral vectors represent one novel strategy to develop specifically designed recombinant vaccines. Many vaccines have to be applied to certain risk groups among the population that represent large parts of the population, such as infants, children, the elderly, or people with a compromised immune system. Because of the altered competence of the immune system of such individuals, the safety issue is of prime importance. In the case of viral vector-based vaccines, safety will be increased if they are rendered fully replication deficient because replication deficiency does not allow vector spreading and persistence within the vaccinees or mutation toward a more pathogenic variant. In the present chapter we first take up, as an example, respiratory syncytial virus (RSV) to discuss the past problems and future options in its vaccine development. We then illustrate our efforts to develop a genome replication-deficient Sendai virus (SeV) vaccine vector. Three essential prerequisites had to be met by such a vector: (1) complete genome replication deficiency while still being able to efficiently express the inserted vaccine antigen genes; (2) stability of viral genome and encoded *trans*-genes; and (3) production of the replication-deficient vector stock to a sufficiently high titer in a *trans*-complementing cell culture system.

Replication deficiency of SeV vector could be achieved via different modifications of components involved in the viral replication complex that consists of the viral N, P, and L proteins. The viral polymerase can operate in two ways: transcription

M. Wiegand (✉)
AmVac Research GmbH, Lochhamer Strasse 29a, 82152 Martinsried, Germany
e-mail: wiegand@amvac.eu

W.J. Neubert
Department of Molecular Virology, Max Planck Institute of Biochemistry,
82152 Martinsried, Germany

Y. Nagai (ed.), *Sendai Virus Vector: Advantages and Applications*,
DOI 10.1007/978-4-431-54556-9_4, © Springer Japan 2013

to enable gene expression and genome replication to generate progeny genomes. The difficulty was to keep the polymerase efficiently performing transcription while its replication activity was completely switched off. Rational design and reverse genetics allowed us to analyze various SeV variants with mutations in the *N*, *P*, and *L* genes. Finally, uncoupling of transcription from replication could be achieved via deletion of an N-terminal part of the P protein that encodes amino acids 2 to 77. Using this vector backbone, the first prototype divalent vaccine against RSV and parainfluenza type-3 virus was developed. The vaccine showed a sufficiently high production titer and high genetic stability. It proved to be fully replication deficient while still expressing the Sendai viral and transgenes at significant levels and was able to elicit a broad spectrum of humoral and cellular immune responses in animals. This preclinical proof-of-concept should lay the ground to use this novel vector platform for specific developments of vaccines with an enhanced safety profile.

4.1 Introduction

Human health is still threatened by widespread infectious diseases against which no proper therapeutic means yet exists. The medication of choice would be a prophylactic solution such as a vaccine to limit the disease burden. However, the classical approach to develop effective vaccines has been unsuccessful for a number of dangerous infectious diseases caused by microorganisms, such as HIV/AIDS, tuberculosis, cholera, malaria, dengue virus, hepatitis C, leishmania, or respiratory syncytial virus (RSV). There is a high unmet need for novel approaches that can trigger an immune response in a specific way which sufficiently stimulates the relevant components of the immune system against each specific pathogen. At the same time, mass medications such as vaccines especially have to meet rigorous safety requirements.

4.2 Safe Viral Vectors for Vaccination

In view of different vaccination populations and vaccine applications, the safety issue should be considered differentially. For example, normal healthy adults who decide to be vaccinated against a pathogen to which they might not be exposed under normal circumstances (e.g., yellow fever vaccine for travelers) might not be comparable to children who receive general vaccinations in early childhood against measles or polio. For such vaccinations, the paramount prerequisite is safety. To meet all safety requirements, novel vaccines based on viral vectors should optimally be characterized as follows:

- No uncontrolled spread of the vaccine: the vaccine should under any circumstances be limited to the site of application without being able to spread within targeted tissues or to unwanted tissues.

- No persistence within vaccinees: the vaccine should not be able to persist within a vaccinee at sites hidden for immune surveillance or by low-level replication.
- No uncontrolled genetic modifications: the vaccine should not mutate in any way in the vaccinee because mutations cannot be controlled and the outcomes cannot be foreseen.

All these requirements can be fulfilled by inactivated (killed) or replication-deficient vaccines. Optimally, these vaccine prerequisites should be combined with high efficacy and long-lasting protection based on a comprehensive immune response comprising specific humoral and cellular responses. Specifically designed vaccination platforms could then be employed for the improvement of existing vaccines but, even more importantly, for the development of novel vaccines against diseases to date unpreventable.

Another aspect of the development of such a novel class of absolutely safe and efficacious vaccines could be an increase in vaccination compliance rates. Especially in some developed countries, where overall levels of health and welfare are high, the public acceptance of vaccination is decreasing more and more. If this trend continues, the socioeconomic impact of infectious diseases will slowly increase, compromising some of the great medical achievements of the past decades and centuries. Excluding any risk through increased safety of novel vaccines will greatly contribute to attracting more people to vaccination.

4.3 Vaccine Development Against Respiratory Syncytial Virus (RSV)

RSV is an example of a worldwide spreading respiratory pathogen, but no vaccine has yet been marketed despite a high medical need.

4.3.1 Health Impact of the Diseases Caused by RSV

RSV predominantly infects airway epithelial cells that line the nose as well as the large and small airways. The majority of RSV infections do not result in a severe illness but only cause moderate upper respiratory tract infections (URTI), resulting in common cold-like symptoms. When RSV causes lower respiratory tract infections (LRTI), they are often manifested as severe bronchiolitis or pneumonia, which can lead to respiratory failure. The virus spreads via respiratory secretions through close contact with contaminated surfaces and objects. The peak incidence of infections is seen in the winter months, usually coinciding with the influenza epidemic. Certain risk groups acquire severe RSV infections at a much higher incidence than normal healthy adults. These risk groups include the elderly, people with underlying diseases, premature infants, children under 4 years of age, and immunocompromised patients.

Table 4.1 Respiratory syncytial virus (RSV) disease burden in seven major markets (Japan, USA, Canada, Germany, France, UK, Italy, Spain). (Adapted from Datamonitor 2006)

Risk group	Total prevalence	RSV infected	Hospitalized because of RSV infection	RSV infected but not hospitalized
Elderly	116,930,000	5,554,175	576,600	8,015,610
Adults with underlying disease	46,739,000	3,038,035		
Children under 4 years of age	41,911,000	9,220,420	258,172	8,962,248
Premature infants	919,000	367,600	91,900	275,700
Bone marrow recipients	18,123	1,812	1,812	0
Total in 7MM	206,517,123	18,182,042	928,484	17,253,558

In total, more than 900,000 people are hospitalized annually because of an RSV infection in the seven major markets (Table 4.1).

The worldwide burden estimated by the WHO is 64 million infections with 160,000 deaths each year. In children less than 5 years of age, more than 30 million infections are attributed to RSV each year with more than 3 million hospitalizations, highlighting the high socioeconomic burden (Nair et al. 2010). Up to 200,000 deaths are estimated to be related to acute LRTI, 99 % of which occur in developing countries. In the United States (US) RSV infections result in up to 150,000 hospitalizations each year, the largest share in children less than 1 year of age (Shay et al. 1999; Zhou et al. 2012). Most children are infected by RSV at the age of 2 (Glezen et al. 1986). Reinfection can occur throughout the lifetime; however, clinical symptoms become milder upon reinfection of fully immunocompetent people. Elderly people represent the second-largest risk group to acute LRTI through RSV with an estimated 15,000 deaths per year (Falsey et al. 2005).

A vaccine against RSV is of urgent need; however, because principally certain risk groups are the target population, the safety issue is crucial.

4.3.2 *Biology of RSV*

RSV is classified in the genus *Pneumovirus* of the family *Paramyxoviridae*. The RSV genome is composed of a single-stranded negative sense RNA of about 15,200 nucleotides (nt) encoding 11 proteins. Seven genes encode the structural proteins N, P, M, F, G, SH, and L and the other four encode the nonstructural proteins NS1, NS2, and M2-1/M2-2 with regulatory functions on the viral life cycle and host immune responses (for review, Collins and Crowe 2007).

There are two known serotypes of human RSV, A and B, circulating only among humans without any relevant genetic alteration to evade the immune response. One strain tends to dominate during an epidemic in an individual location, although at times both strains can be isolated from patients in the same area. In the US and UK strain A is found more commonly, whereas strain B appears more frequently in continental Europe. Although the two strains can be discriminated in diagnosis, the antigenic diversity of both strains is rather low. Nonhuman RSV types also exist, such as bovine RSV; however, bovine RSV exhibits only limited sequence homology with the human strains, inducing no efficient cross-protection against human strains.

4.3.3 RSV Vaccine Development

Human RSV was discovered in the late 1950s, and RSV vaccine development was initiated in the 1960s. The initial product, formalin-inactivated (FI) whole virus particles, was efficacious. However, it was subsequently found that this FI-RSV vaccine caused a more severe disease upon natural RSV infection in the group of immunized children compared to the control group (Kapikian et al. 1969), even resulting in two fatalities. This disaster ignited a long period of basic research to understand the disease pathology and host–pathogen interactions and to create safer and efficacious vaccine candidates for future use.

The observed disease enhancement by FI vaccine was associated with pulmonary pathology caused to a large extent by infiltrating eosinophilia attracted by a full-blown RSV infection. Obviously, the immune response to the FI-RSV was incomplete or not stimulating relevant immune responses that could contain and ultimately clear the infection. This response was subsequently found to be associated with the induction of an imbalanced Th2-biased T-cell response after RSV infection in different FI-RSV-vaccinated animal species (Connors et al. 1994; Waris et al. 1996; de Swart et al. 2002).

4.3.4 RSV Vaccine Candidates

4.3.4.1 Live-Attenuated Vaccines

There is a long history of attempts to develop live-attenuated RSV vaccines. In principle, live-attenuated vaccines are expected to elicit the broadest, and thus most protective, immune response. The reason why no live-attenuated vaccine has yet been marketed is again related to the difficulties specific to RSV. Certain risk groups show compromised immune functions, complicating the search for the right balance between over-attenuation, which results in minimal efficacy, and under-attenuation, causing disease or severe side effects. The first and follow-up generations of live-attenuated RSV vaccines that were created acquired amino acid changes through serial virus passages in cell culture under nonnatural growth conditions, such as at cold temperatures (Pringle et al. 1993; Karron et al. 1997; Wright et al. 2000).

After reverse genetic techniques became common laboratory praxis in the mid-1990s, more specifically attenuated vaccine candidates were created by introducing predetermined attenuating mutations into the RSV genome. For example, genes encoding the nonstructural proteins that interfere with the host immune response were deleted to develop less harmful vaccine candidates (Jin et al. 2000).

The first clinical trials showed that live-attenuated RSV vaccines could be applied safely without causing any enhanced disease for the vaccinees at risk (Karron et al. 2005). This important observation prompted researchers to broaden the search for an efficient live RSV vaccine. The clinically most advanced candidate to date is a temperature-sensitive cold-adapted RSV with several mutated amino acids combined with a deletion of the *SH* gene (MEDI-559; MedImmune). Tolerability, viral shedding, genetic stability, and immunogenicity are currently being evaluated in a clinical phase I/IIa study for children 1 to 24 months of age. Another attenuated candidate vaccine currently under safety and immunogenicity evaluation in a clinical phase I with seropositive and seronegative children is a recombinant RSV with the *ΔM2-2* and *NS1* genes deleted (Jin et al. 2000; Wright et al. 2006).

4.3.4.2 Viral Vectors

Live viral vector vaccines represent another very promising approach for the development of a safe and efficient RSV vaccine. Different vector systems have been used as a platform but are mostly still restricted to preclinical evaluation.

The most advanced candidate is based on a chimeric bovine/human parainfluenza virus that additionally expresses the human RSV F protein. Attenuation is related to this mixed phenotype of human envelope proteins (F and HN) and remaining bovine proteins. This candidate already showed safety for seropositive children 1 to 9 years of age (Gomez et al. 2009). Also, Sendai virus (SeV) has been used in a clinical phase I study (Slobod et al. 2004). In preparation for its future application as a safe vaccine platform, SeV was administered to healthy human adults, however, first as Jennerian vaccine against its human counterpart parainfluenza virus type 1 (PIV1). Safety and tolerability could be shown. Encouraged by this outcome, a recombinant SeV vector expressing the RSV F protein was created as an RSV vaccine candidate. Its preclinical efficacy was shown in cotton rats (Zhan et al. 2007).

Adenoviral vectors have also been in use for a long time, and different RSV vaccine candidates have been tested preclinically so far. Single intranasal immunization with a recombinant adenovirus expressing the core unit of the RSV G protein in three tandem repeats induced protection in mice (Yu et al. 2008). Another replication-deficient adenovirus vector expressing the RSV F protein protected intranasally immunized mice from RSV challenge infection with a well-balanced T-helper cell response, suggesting the absence of vaccine-enhanced disease (Fu et al. 2009).

Poxviruses have also been used as an RSV vaccine vector by different research groups. A replication-deficient MVA (modified vaccinia virus Ankara) that expressed both the RSV G and F protein showed preclinical efficacy in an intranasal prime and intramuscular boost scenario (Wyatt et al. 1999). Surprisingly, however, no protection was found when subsequently applied to infant macaques (de Waal et al. 2004). In another approach an MVA-based RSV vaccine candidate showed preclinical efficacy, again, however at the same time caused some risk of inducing RSV vaccine-enhanced disease (Olszewska et al. 2004).

4.3.4.3 Viral Replicon Particles (VRP)

Another vaccine approach is adapted from the afore-presented viral vector approaches and uses only the replicative units of RNA viruses to express the antigenic component(s). These so-called replicon particles cannot produce infectious progeny particles in target cells because the structural protein genes are removed from the entire viral genome and replaced with heterologous antigen genes. In this way, a high level of safety will be guaranteed. However, because of the presence of all viral components required for replication, the VRP is able to self-amplify in cells. Whether this causes any risk of persistent infections remains to be seen. So far two different VRPs expressing the RSV F protein were developed based upon alphaviruses, Venezuelan equine encephalitis virus, and Semliki Forest virus with a single-stranded, positive sense RNA genome. The induction of robust cellular and humoral responses as well as mucosal responses has been associated with their proven protection capacity against RSV challenge in a preclinical setting (Mok et al. 2007; Geall et al. 2012).

4.3.4.4 Subunit Vaccines

Another promising approach to maximize the safety is to use subunit vaccines. Different RSV subunit vaccine candidates have been developed in the past based on either purified RSV F protein, the G protein fused to an alum-binding domain, or a mixture of RSV M, F, and G proteins. These products were tested clinically in healthy adults or elderly people with or without coadministered alum as adjuvant. Overall, some efficacy was found but the rise in neutralizing antibody titers was only modest (fourfold or less) and limited to few vaccinees (10–50 %) (Falsey and Walsh 1996; Power et al. 2001; Piedra et al. 2003; Langley et al. 2009).

4.3.4.5 Virus-Like Particles (VLP)

A variant of subunit vaccine is virus-like particles (VLP) that express the vaccine antigens, the RSV G or F proteins, on their surface. Again, safety of this approach is a clear advantage over live vaccines, so long as efficacy can be proven. Several different VLP systems showed protection against RSV challenge in a preclinical animal model (Murawski et al. 2010; Stegmann et al. 2010; Quan et al. 2011; Smith et al. 2012). The RSV VLPs were applied systemically with or without coadministered adjuvant. One approach showed safety and immunogenicity in a clinical phase I setting with healthy adults (Glenn et al. 2013) and is currently being examined in a phase II study with healthy women of childbearing potential.

4.3.4.6 DNA Vaccines

More recently, nucleic acid-based vaccines encoding the antigenic components of RSV also have been developed. A codon-optimized RSV F protein-expressing plasmid DNA showed some protective efficacy in mice after two subcutaneous applications of 50 μg DNA (Ternette et al. 2007). In another study calves were immunized with plasmid vectors encoding the RSV *F* and *N* genes. Interestingly, a strong cell-mediated specific immune response was developed and associated with protection (Boxus et al. 2007). Sufficient delivery of plasmid DNA to the nucleus of target cells as well as safety of application are probably the main issues to be solved for this approach.

4.3.4.7 Passive Immunization

All the aforementioned approaches have clear advantages that support their use, although efficacy has not yet been established for most of them in a clinical setting. The only marketed product related specifically to prevention of RSV-related disease is a monoclonal antibody (Palivizumab) directed against the RSV F protein. This product, however, is only applied to passively immunize high-risk newborns because of its high costs (administration of 15 mg/kg body weight on a monthly basis) and its limited efficacy (the absolute risk reduction of RSV-related hospitalization is 5.8 %) (The IMpact-RSV Study Group 1998; Cardenas et al. 2005). Although this medication is restricted to a very limited number of individuals, it has demonstrated that a severe RSV pathology during infection can be reduced or prevented through the administration of neutralizing antibodies against the RSV F protein. Use of monoclonal antibodies as a therapeutic agent for severe RSV infection is spreading in developed countries.

4.3.5 Requirements for RSV Vaccine Development

No RSV vaccine for active immunization has yet been introduced into the market. What are the main factors that complicate the development of an efficient RSV vaccine on the one hand, and on the other, what is the optimal profile of an RSV vaccine?

The two main components required for an efficient vaccine are the safety and efficacy. They have to be dissected specifically for RSV vaccines. As already described, FI-RSV caused enhanced disease upon natural RSV infection, likely resulting from a Th2-biased response of helper T cells (Openshaw et al. 2001; de Swart et al. 2002). Therefore, most important in terms of safety of an RSV vaccine is not to induce disease enhancement. Polarization analysis of the T-helper cell response upon vaccination and subsequent challenge as well as histopathological analysis are mandatory in the process of identifying promising vaccine candidates. The choice of the RSV antigen can have a huge impact on the T-helper cell response. When subunit or viral vector vaccines are applied, any RSV proteins can be selected as antigens. In principle, the envelope proteins are well suited to induce neutralizing antibody responses. Therefore, for RSV, the G and the F proteins are regarded as main targets. The G protein, however, and especially its secreted form, which appears to be synthesized in a certain proportion during gene expression, was shown to be associated with the induction of a Th2-biased response and enhanced illness upon RSV challenge in mice (Johnson and Graham 1999; Bukreyev et al. 2008; Moore et al. 2008).

The safety aspect related to an RSV vaccine has to be even more pronounced because various high-risk groups are the target population for vaccination. Inherent to these risk groups is that their immune competence is restricted or limited to a certain extent, either because it is not yet fully mature—as with infants or newborns—or compromised because of an immune defect, immune suppression (e.g., transplant recipients), or aging. The course of an immunization leading to sufficient immune response therefore has to be highly predictable without any clear pharmacodynamics. Thus, if live-attenuated or vector-based vaccines are applied, it has to be assured that no uncontrolled spread of the vaccine can occur in those vaccinated risk groups under any circumstances. Therefore, live-attenuated or live vector-based RSV vaccines have to be attenuated sufficiently to exclude any risk. In addition, the introduced attenuating mutations have to be maintained stably during vaccination. In this respect, there has to be a clear idea of the number of replication cycles that a vaccine undergoes in the vaccinee. Taking this into account, the risk of a potential reversion or further mutations with unforeseen outcomes could be assessed. However, when following the approach of killed or replication-defective vaccines, this risk can be minimized.

Many factors are crucially involved in RSV vaccine efficacy. Efficacy is mainly associated with the induction of sufficient neutralizing antibodies as with most other vaccines. The only marketed prophylactic or therapeutic product against RSV so far, the monoclonal antibody palivizumab, has demonstrated neutralizing efficacy against the RSV F protein. At the same time, the antigenic component should stimulate protective responses across the two RSV serotypes A and B. Induction of cross-neutralizing antibodies was possible because of the significant sequence conservation between the F proteins of two serotypes (Collins and Crowe 2007; Anderson et al. 1985).

The CD8(+) cytotoxic T lymphocytes (CTL) play a major role in the clearance of RSV-infected cells and, thus, have a huge impact on the clinical outcome of RSV infection and vaccine development (Alwan et al. 1992). For the RSV G protein it was shown to lack MHC I-restricted epitopes (Srikiatkhachorn and Braciale 1997) and, thus, cannot elicit CTL response (Heidema et al. 2004). RSV F, in contrast, contains a major CTL epitope in mice and humans (Levely et al. 1991; Cherrie et al. 1992; van Bleek et al. 2003). The SeV vector expressing the RSV F protein induced a T-cell response that was protective in the absence of specific antibodies (Voges et al. 2007).

Another important factor contributing to the efficacy of vaccines against respiratory viruses is mucosal immunity. Pathogens can be attacked directly upon the first encounter and manifestation or spread of the infection can be drastically reduced or, at best, completely inhibited. The mucosal immune response is also considered to be essential for an effective RSV vaccine. RSV infection is restricted to the epithelial tissue of the respiratory tract. Although serum antibodies are likely sufficient to protect the lower respiratory tract (LRT) because of a low gradient between blood and tissue, the upper respiratory tract (URT) needs support from local immune responses such as secretory IgA. Serum antibody level does not suffice to prevent RSV (re-)infection at the URT. An efficient local mucosal immune response could be elicited through application of the vaccine via the mucosal route, if applicable to the respective vaccine approach.

Efficacy of a vaccine can be compromised by preexisting immunity against the vaccine. In this context, two factors should be taken into consideration for RSV vaccines: the target population and the nature of the vaccine itself. There are different potential target populations for an RSV vaccine. All populations except children less than 1 year of age have had (repeated) contact with RSV during their lifetime and consequently have some level of preimmunity, even though not protective. Therefore, approaches such as live-attenuated viruses or viral vectors can be compromised by preexisting antibodies against them, resulting in reduced efficiency. Optimally, pre-immunity is absent against the vaccine delivery platform as, for example, with synthetic particles, some subunit vaccines, or nonhuman vectors. For vaccination of seronegative infants, pre-immunity could only play a role if a significant maternal anti-RSV antibody level is present during the first months of life.

4.4 SeV Vector Platform

SeV has evolved into a novel promising vector platform during the past two decades. Active incorporation of basic knowledge on SeV biology facilitated this evolution.

4.4.1 SeV Biology

SeV belongs to the genus *Respirovirus* in the family *Paramyxoviridae*. The genome of 15,384 nt encodes six structural proteins (N, P, M, F, HN, L) from separate genes and several nonstructural proteins (C′, C, Y1, Y2, V, W, X) from alternative reading frames or template variation within the *P* gene. Via its surface protein HN (hemagglutinin-neuraminidase), SeV attaches to the ubiquitous receptor structures containing sialic acid. The viral F protein subsequently mediates fusion of the viral envelope with the cell membrane, and the ribonucleoprotein (RNP) is released into the cytoplasm, where the whole intracellular viral lifecycle takes place. The viral RNA-dependent RNA polymerase is able to operate in two modes: either as the transcriptase to generate capped and polyadenylated mRNAs that are translated into viral proteins, or as the replicase to generate full-length, nucleoprotein (N)-encapsidated genomes and antigenomes. For details, see Chap. 2.

In the beginning of infection, transcription is emphasized to synthesize first a substantial amount of viral proteins to support viral replication. During this process, transcription enters different productive stages. The very first transcription, termed early primary transcription, is completely dependent on the activity of viral polymerase molecules that entered the cell being associated with the infecting encapsidated viral genome. The polymerase molecules create initial mRNAs directing the viral proteins. Among those proteins, newly synthesized P and L proteins constitute new polymerase molecules that further support the viral transcription. This second phase of transcription is called late primary transcription (Wiegand et al. 2007; Bossow et al. 2012) (also see Chap. 2). After sufficient N protein has been produced, the viral polymerase also starts to synthesize new antigenomes and genomes, which are full-length RNA transcripts that are concomitantly encapsidated with viral N proteins, each N protein covering six nucleotides. After new genomic templates have been generated, transcription enters the secondary stage with enhanced productivity as a result of the growing number of available templates (also see Chap. 2).

4.4.2 Prerequisites and Requirements for a Safe SeV Vector

Here we describe the development of an SeV vector intended to combine efficacy of live vectors with high safety. To achieve this goal we followed the approach to generate a replication-deficient vector that is still capable of efficient gene expression including the expression of heterologous transgenes. The following aspects summarize the potential advantages of such a vaccine vector.

4.4.2.1 General Features Intrinsic to SeV Vaccine Vector

Pathogenicity

SeV is nonpathogenic to humans, even though it can infect human tissues in general. The risk to cause a disease in the vaccinee is regarded as very low.

Preimmunity

The natural hosts of SeV are rodents. SeV-based vaccine approaches are not expected to be faced with antivector pre-immunity that might decrease efficacy, as already demonstrated for different SeV vector-based applications (e.g., see Chap. 8; Moriya et al. 2008, 2011). Antigenic cross-reactivity with human counterpart PIV1 is an unsettled problem because anti-PIV1 is widely prevalent in humans and may counteract the infecting SeV. So far, however, the anti-SeV titers in human sera were quite low. Repeated administration of SeV with an appropriate interval was possible in a monkey infection model. For details, see Chap. 5.

Mucosal Application

SeV is known to infect mucosal epithelial cells very rapidly and efficiently (Yonemitsu et al. 2000; Griesenbach et al. 2002; also see Chap. 3). Therefore, SeV vector vaccines are very well suited for mucosal application and induction of mucosal immunity.

Broad Host Range but Restricted Tissue Tropism

SeV attaches to host cells via ubiquitously present sialic acids containing surface structures. Naturally SeV is restricted to growth in epithelial cells of the respiratory tract of rodents because of a specific protease occurring only in this tissue. This protease activates the F protein by cleaving the inactive F_0 precursor protein into the active subunits F_1 and F_2 that remain disulfide bonded. Thus, although SeV can infect nearly all cell types, regardless of actively propagating or quiescent, its spread is limited to the respiratory tracts of rodents. For details see Chap. 2.

These and some other properties unique to SeV including nonintegrating nature (into host cell chromosomes) are thought to be the advantages of the SeV vector (see Chap. 3).

4.4.2.2 Necessity of Generating Replication-Deficient SeV

No Uncontrolled Spread

Genome replication deficiency could exclude the opportunity of any uncontrolled virus spread in the vaccinee because of the absence of newly synthesized genomes and hence no infectious progeny production in infected cells.

Genetic Stability, No Reversion, and No Unintended Mutations

Genetic instability, reversion to the wild type, and unintended spontaneous nucleotide changes are always the issues of high concern in vaccine products. These changes occur only during replication and accumulate in the replicated genome but not in the original template. Then, if the live-attenuated viruses or live vectors are completely replication incompetent, the vaccinees need not be afraid of those unforeseen adverse events. Thus, the replication-incompetent constructs that still are capable of vaccine antigen expression represent an ultimately safe and efficacious tool.

No Persistence

The vaccine delivers functional nucleocapsid template into the target cells where gene expression goes on as long as protein degradation terminates the functional integrity of the template. Once degradation of the N proteins attached to the viral RNA genome starts, the RNA genome becomes unprotected and will be rapidly digested by cellular nucleases. This stage eliminates the risk of any persistence of the vector.

4.4.3 Generation of a Replication-Deficient SeV Vector

We initiated the attempts to find an optimal way to render SeV vector replication deficient without dismantling its primary transcriptional capacity. The viral components directly involved in transcription and replication are the members of the viral nucleocapsid: (1) the RNA genome encapsidated with the N proteins (= the ribonucleoprotein, RNP); (2) the viral polymerase consisting of the catalytic unit, the L protein, and the cofactor, phosphoprotein P; the latter is present as a homo-tetramer (P_4) within the P_4–L complex; and (3) the accessory C protein that is likely responsible for determining the negative polarity of genome (for details, see Chap. 2). The remaining gene products such as M, F, and HN proteins are not regarded as valid targets to render SeV replication deficient. SeVs with those genes deleted potentially replicate and persist in the target cells or even produce progeny particles (see Chap. 3), and hence do not meet the highest safety requirement, replication incompetence.

In the 1990s, two different approaches became available to study SeV RNA synthesis. One is a cell culture-based system wherein the viral L, P, and N proteins are expressed and their roles are assessed on the transcription and replication of cotransfected minigenome (a small defective interfering RNA genome named DI-H) (Curran et al. 1991). In this way, all the three proteins (L, P, and N) were demonstrated to be essential for SeV RNA synthesis. However, it was not easy to dissect transcription from replication. The other approach was to reconstitute the RNA synthesis steps *in vitro* by incubating purified genome templates and the extracts from infected/cotransfected cells (Horikami et al. 1992). Using this setup, Curran et al. (1992) revealed that providing the P–L complex is sufficient and no newly synthesized N protein is needed for transcription (mRNA synthesis). For viral genomic replication, instead, two distinct protein complexes were required: the P–L complex forming the active viral polymerase and another complex consisting of N and P proteins (N^0–P) that mediates encapsidatation of the newly synthesized genomic RNA with newly synthesized monomeric N proteins (Horikami et al. 1992) (Table 4.2).

Table 4.2 Requirement of protein complexes for viral transcription and replication

	Transcription	Replication
N–P complex	–	+
P–L complex	+	+

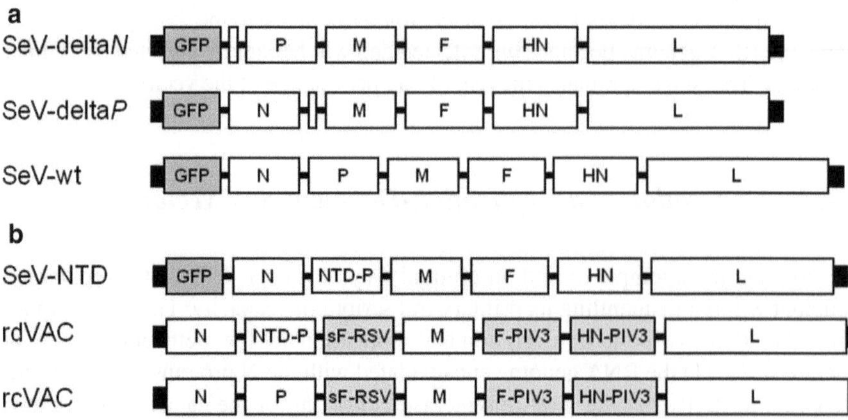

Fig. 4.1 Schematic representation of the vector genomes of (**a**) SeV-delta*N*, SeV-delta*P*, and SeV-wt and (**b**) of SeV-NTD, rdVAC, and rcVAC. SeV-delta*N* and SeV-delta*P* had full deletions of the *N* or *P* ORFs, respectively. SeV-delta*N*, SeV-delta*P*, SeV-wt, and SeV-NTD had a *GFP* transgene inserted behind the leader. SeV-NTD and rdVAC had an N-terminal deletion within the *P* gene

4.4.3.1 Replication Deficiency via the Deletion of the Entire *N* Gene or the Entire *P* ORF

Either the L or P or N protein can be the target of manipulation to create replication-deficient SeV (Table 4.2). We succeeded in generating SeV vectors with the entire *N* gene deleted or with the entire *P* ORF deleted, named SeV-delta*N* and SeV-delta*P*, respectively (Fig. 4.1a). As a marker gene the green fluorescent protein (GFP)-encoding *trans*-gene was inserted behind the leader sequence. The wild-type (wt) SeV vector also encoding the GFP gene served as a fully replication- and transcription-competent control vector (Fig. 4.1a).

To produce SeV-delta*N* and SeV-delta*P* particles in sufficient amounts, the missing proteins were in *trans* provided from an external source. A helper cell line generated through stable transfection with the *N* gene would be the solution of choice to provide SeV-*deltaN* with the missing N protein. However, the N protein is needed in a large amount. Actually, as many as 2,564 units of N protein are needed to finish the encapsidation of a single copy of SeV wt genome or antigenome of 15,384 nt because a single N covers 6 nt and, thus, has to be provided in quite a large quantity by the helper cell line to facilitate efficient viral replication to obtain sufficient vector stock. On the other hand, the number of P and L proteins per virion was estimated to be 300 and 50, respectively (Lamb et al. 1976). In addition, the N protein encapsidates RNA illegitimately in cells without the concomitant presence of the P protein (Buchholz et al. 1993).

The first generation of helper cells, H29, based on HEK-293 cells were generated that expressed SeV N, P, and L proteins simultaneously (Willenbrink and Neubert 1994). The amount of N and P proteins expressed there was 5 % and 10 %, respectively, compared to those present in SeV wild-type (wt)-infected cell. In this

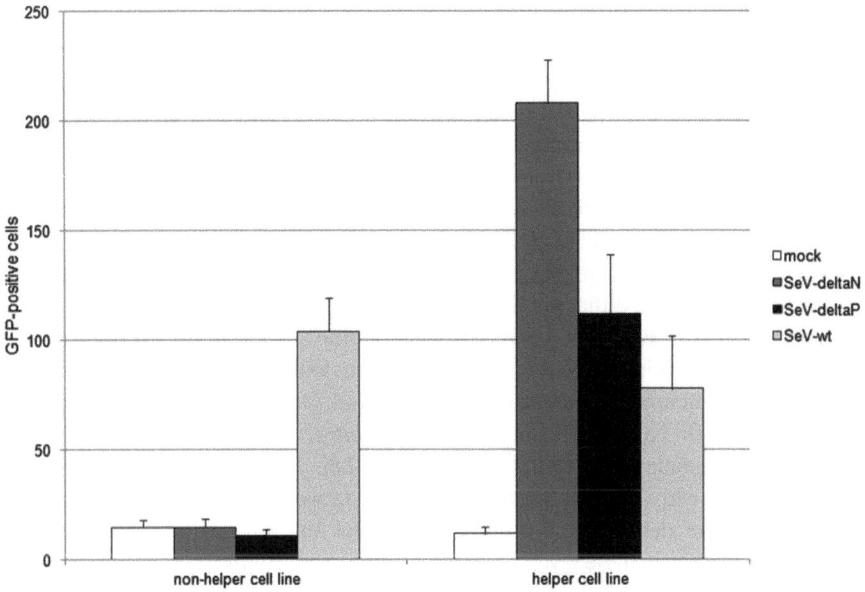

Fig. 4.2 Gene expression analysis of SeV-delta*N* vector in helper and non-helper cell lines. *GFP* transgenes were encoded on SeV-delta*N*, SeV-delta*P*, and SeV-wt. Helper and non-helper cells were infected with MOI 0.05 and analyzed by FACS for GFP expression 3 days post infection (p.i.)

helper cell line, DI-H particles (genome size, 1,410 nt) that do not encode any full-length viral protein could be successfully replicated (Willenbrink and Neubert 1994). After this proof-of-concept was shown using DI-H particles, it was still questionable whether these quantities of proteins would be sufficient to efficiently *trans*-complement SeV-delta*N* or SeV-delta*P* particles. In an attempt to increase the amount of expressed N and P proteins, additional helper cell lines were generated, without, however, significantly raising the protein levels.

After generation of SeV-delta*N* particles via virus rescue—a widely used and well-established process—they were amplified in the helper cells. The same cells could be used to amplify SeV-delta*P* particles obtained from another virus rescue. Remarkably, the amount of produced SeV-delta*P* clearly exceeded that of SeV-deltaN. Obviously, the amount of N protein required to fully *trans*-complement SeV-delta*N* could be too high to be reached in a stable helper cell line, possibly because of the cytotoxic nature of the N.

Even though the overall yield of SeV-delta*N* was low, the amount of particles was sufficient to be used for test infections in cell culture. In a non-helper cell line, no gene expression, as monitored by GFP expression, was detected not only for SeV-delta*N* but also for SeV-delta*P*, in contrast to SeV-wt, which expressed GFP vigorously (Fig. 4.2) (Wiegand et al. 2007). This outcome was very astonishing and had not been expected based on the *in vitro* results reported earlier (Horikami et al. 1992; Curran et al. 1993).

In conclusion, engineering of SeV genome toward this direction, that is, deletion of the entire *N* gene or the entire *P* ORF, did not appear to reach our goal to create a replication-inert and transcription-active vector. However, these constructs, along with other constructs of partial deletion of the *P* ORF, facilitated greatly our understanding of SeV primary transcription, proposing a novel concept that the primary transcription starts with an extremely low, basal level in the preliminary (early) phase, and then proceeds into a late phase of normal level by the help of *de novo* synthesized N and P proteins (Wiegand et al. 2007; Bossow et al. 2012). The genome replication can start only after this late phase. (Also see Chap. 2 for some details.) These novel results clearly indicated that not all results obtained by prior *in vitro* analysis can be reproduced *in vivo*. Based on the experimental setting and the afore-described results, this is not very surprising; the prerequisites inherent to the *in vitro* system were in part substantially different from the real *in vivo* situation. One or a few NCs are employed *in vivo* versus 10^{11} NCs (Horikami et al. 1992) *in vitro*, which does not reflect the situation at the beginning of a natural infection; further, discrimination of transcription activity during early and late primary transcription was not possible by the *in vitro* system. However, detecting this difference is crucial for the development of a efficient replication-incompetent SeV vector. Also, the importance of the presence of N protein for transcription could not be detected using the less-sensitive *in vitro* setup.

4.4.3.2 Replication Deficiency via Modification of the *L* Gene

The *L* gene represents the largest gene (nearly 7 kb), accounting for nearly half of the 15.4-kb SeV genome. The translation product is composed of 2,228 amino acids and is regarded to harbor all catalytic active sites of the viral RNA-dependent RNA polymerase (Hammond and Lesnaw 1987; Hercyk et al. 1988; Einberger et al. 1990; Poch et al. 1990; Hammond et al. 1992; Hunt and Hutchinson 1993). Via comparison of sequences of L proteins among different nonsegmented negative strand RNA viruses, six conserved domains were identified, which were termed I–VI (Poch et al. 1990; Sidhu et al. 1993; also see Chap. 2). Based on *in vitro* mutation analyses of different parts in these domains, single amino acids and regions were identified that exhibited great influence on transcription and replication functionality.

Within domains II and III, highly conserved regions were found for SeV and other nonsegmented negative strand RNA viruses. Single amino acid mutations within these regions resulted in abrogation of any RNA synthesis (Jin and Elliott 1992; Smallwood et al. 1999; Sleat and Banerjee 1993; Schnell and Conzelmann 1995). Some other mutants showed reduced genome replication activity compared to transcription (Smallwood et al. 1999; Smallwood et al. 2002a; Smallwood et al. 2002b). *In vitro* some of the analyzed mutants, mostly carrying single amino acid substitutions, showed transcription being uncoupled from replication. One special mutant did not show any remaining replication activity while transcription levels were still at roughly 50 % of the wild-type level. Some mutants showed different results of their capacity of replication when analyzed by *in vivo* versus *in vitro* assays. Studies can monitor only a single round of replication whereas multiple rounds of replication can be seen *in vivo*. However, still no uniform picture could be depicted; sometimes *in vitro* replication, while in other cases *in vivo* replication, was pronounced. Unfortunately, the capacity of these mutants to perform transcription was very poor. Only a few mutants with reduced or fully abolished genome replication behavior *in vitro* or *in vivo* exhibited transcription capacity, and at very low levels. So far, no clear-cut dissection of transcription from replication appeared to be possible. Mutational studies in domains IV and VI identified two mutants for which replication was inert but transcription active (Feller et al. 2000). Here again, however, the transcription rate was significantly lowered.

Overall, it appeared quite difficult to generate a completely replication-inert and sufficiently transcription-active SeV based upon those previously identified mutations. From a technical perspective, the large size of the *L* gene complicates research compared to smaller genes such as the *N* and *P* gene, with respect to cloning procedures or sequence monitoring, for example. In addition, the L protein is very complex and functions as a multi-activity enzyme. The L interacts with many different components of the virus as well as host cell factors. In this way, small modifications can easily result in unforeseen activities or functionalities. Measuring and controlling these can be very cumbersome and complex.

4.4.3.3 Replication Deficiency via the N-Terminal Deletion of the *P* ORF: a Strategy

The *P* gene is organized in a more complex way compared to the other SeV genes: it is polycistronic. Besides the structural protein P, it encodes several nonstructural or accessory proteins from smaller overlapping and alternative reading frames. Figure 4.3 illustrates the different reading frames and the respective encoded proteins. Besides the full-length P protein (568 aa), two shortened proteins, V (384 aa) and W (318 aa), are synthesized starting from the same reading frame as P, and thus sharing the same N-terminus but containing alternative C-termini as a result of the frameshift caused by the so-called co-transcriptional RNA editing to insert one or two G residues after codon 317 (Vidal et al. 1990). A small protein, X (94 aa), is encoded from the same P reading frame but is initiated about 1,500 aa downstream, being thus identical in sequence to the C-terminus of the P protein. In addition, four nonstructural or accessory proteins (C′, C, Y1, and Y2), collectively referred to as the C proteins, are encoded on an alternative reading frame (+1) (Patwardhan and Gupta 1988; Latorre et al. 1998). These proteins share the same C-terminus, but start at different positions: C′ (215 aa), C (205 aa), Y1 (181 aa), and Y2 (175 aa). For details, see Chap. 2.

These structural and nonstructural proteins exhibit the respective specific functions in the viral lifecycle. The full-length P protein binds the newly synthesized soluble N protein (N^0), forming the N^0–P complex in its monomeric conformation, and makes the N protein available as the substrate for encapsidation of the nascent chain of the viral genome. In its tetrameric conformation, the P protein (P_4) builds the viral RNA polymerase (P_4–L complex). The X protein only contains the

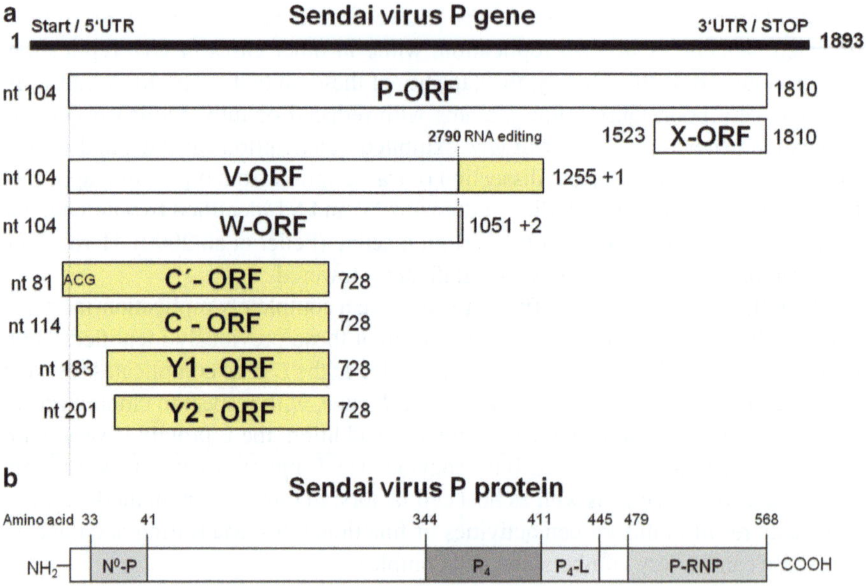

Fig. 4.3 Schematic representation of (**a**) the genomic organization of the *P* gene and (**b**) the functional domains of the *P* gene: domain "N^0–P" contains highly conserved amino acids from position 33 to 41 that bind to the Sendai virus N protein to build the N^0–P complex (Curran et al. 1995b); P oligomerizes via coiled-coil structures within the domain of amino acids 344–411 (Curran et al. 1995a); the "P_4–L" domain harbors amino acids involved in binding of P to the L protein to form the polymerase complex (Smallwood et al. 1994); the "P–RNP" domain is responsible for interaction of the P protein, within the polymerase complex, with the ribonucleoprotein (RNP) (Ryan et al. 1991; Curran et al. 1995a)

functional domain of the P protein responsible for binding to the RNP; however, its explicit function still remains enigmatic. The V not only supports the regulation of viral transcription and replication (Horikami et al. 1996; Yu et al. 1997) but also encodes a luxury function required for pathogenesis to cause severe pneumonia in mice (Kato et al. 1997a, b). The C proteins are mainly associated with evading host response by antagonizing the interferon (IFN)-mediated signaling to induce an antiviral state (Garcin et al. 1999; Gotoh et al. 1999). In addition, the C proteins are reported to play a role in regulating viral RNA synthesis (Curran et al. 1992; Horikami et al. 1997; Latorre et al. 1998; also see Chap. 2).

Because of the complexity of the *P* gene, its modifications aiming at the generation of a replication-deficient viral vector have to be carefully assessed. Full-length deletion of the *P* gene does not result in a viral vector capable of efficiently performing gene expression (Wiegand et al. 2007; see Sect. 4.4.3.1) and therefore does not fulfill the set requirements. In addition, deletions of codons in many parts of the *P* gene lead to deletions in other *P* gene-encoded proteins than the full-length P at the same time, for example, V, W, or C proteins. Any potential impact on replication or transcription behavior of such mutants has to be taken into account and potential functional differences had to be tested.

The full-length P protein has been analyzed mainly through *in vitro* assays in test tubes, and different functional domains have been identified and mapped (Fig. 4.3b). Figure 4.3 depicts domains and assigned functions for the P protein *in vitro*, although they have not yet been fully verified in the natural life cycle *in vivo*. In the analysis of Sendai virus N and L proteins, the results obtained with *in vitro* and *in vivo* experiments have sometimes been contradictory. The questions that we have wanted to address are whether partially deleted *P*-gene mutants are still transcriptionally active, while being replication deficient at the same time, and if so, to what extent.

It appeared to be well established, however, that the N-terminal region (residues 33–41) is responsible for binding soluble, monomeric N^0 to form the N^0–P complex (Curran et al. 1994; Curran et al. 1995b). Without building the N^0–P complex, no encapsidation of newly synthesized RNA genomes and no genome replication would occur. The non-encapsidated, naked RNA would be highly unstable and rapidly degraded by intracellular nucleases. Thus, the deletion of N-terminal region containing the N^0-binding site, leaving other functional sites mapped in the C-terminal half, could be the most plausible strategy to generate replication-inert and transcription-active SeV vector. However, the N-terminal region of P also encodes parts of the six accessory proteins V, W, C′, C, Y1, and Y2. These proteins were reported to have an impact on regulating and influencing viral transcription and replication. Possible consequences of deleting parts of these genes were predicted.

Regarding the C proteins, they were found to have a capacity to down-modulate SeV mRNA synthesis (Curran et al. 1992) and RNA replication selectively via the genomic promoter but not the antigenomic promoter (Cadd et al. 1996; Horikami et al. 1997; Tapparel et al. 1997). Therefore, the mRNA synthesis level appeared to be quite robust in cells infected with a recombinant SeV, 4C(−), that expressed none of the four C proteins, compared with the wild-type SeV infection (Hasan et al. 2000). Also, in 4C(−)-infected cells and in 4C(−) virions, the antigenome was dominant (Irie et al. 2007; for details, see Chap. 2). In this context, the elimination of the four *C* ORFs as a logical consequence of the N-terminal deletion of the P protein could lead to an "enhanced" or "maximum" level of the viral and transgene expression under the conditions of complete abrogation of genome replication, and hence appears to be rather advantageous for the current purpose. However, the loss of anti-interferon capacity caused by the lack of C proteins may be disadvantageous in evading innate host immunity and hence continued expression of the viral genes and transgenes *in vivo*. The effect of the resultant deletion of the N-terminal V protein (a part of the P/V common ORF) remains to be evaluated. To be noted are that the major function of V protein is to maintain a high viral load in mice (Kato et al. 1997a, b) and that this role appeared to be encoded mainly by the cysteine-rich V-unique region (see Chap. 2).

In view of these considerations, we decided to introduce a deletion within the N-terminal part of the *P* gene, which was sufficiently large to ensure the safety of the resulting vector so that spontaneous mutations and reversion to the wild-type would be minimized. This N-terminal deletion (nt 4 to 231) (NTD) resulted in the removal of 76 amino acids (residues 2 to 77) from the *P* ORF. This NTD mutant was formerly named *P*-delta2-77 (Bossow et al. 2012).

4.4.4 Construction and Characterization of the P SeV NTD Vector

We addressed the following three questions:

1. Can such a viral vector with an N-terminal truncated *P* gene be initially generated by reverse genetics und subsequently be produced in sufficient amount using a helper cell system?
2. Does this N-terminal deletion of the P protein fully uncouple transcription from replication *in vivo*?
3. Is the viral transcription and gene expression of this potentially replication-deficient viral vector sufficiently high in stimulating a substantial immune response against the encoded pathogen antigens?

To answer these questions, we followed a stepwise approach using different vector constructs. For the first basic analysis of the key requirements, we generated a viral vector, SeV-NTD, with the N-terminal deletion (NTD) of the *P* gene and, in addition, the GFP marker gene behind the leader (Fig. 4.1b). The production of this NTD mutant in a helper cell system and the viral replication and transcription behavior were studied in cell culture (Bossow et al. 2012).

In the second step, we generated a replication-defective (rd) prototype vaccine vector, rdVAC, that encodes antigens from two viral pathogens in a special way: the soluble form of the F protein of RSV was inserted as an additional transgene behind the *P* gene (Fig. 4.1b) and can serve as an antigen only after being expressed in infected cells. Further, the F and HN surface proteins of SeV were replaced by their counterparts from the human parainfluenza virus type-3 (PIV3) (Fig. 4.1b); in this way, the PIV3 antigens are expected to be expressed in infected cells and finally on the virion surface as the structural components. With this strategy the impact of the multivalent antigen presentation from NTD mutant on immunological responses would be evaluated *in vivo* after immunization of mice (Wiegand et al. 2013).

4.4.4.1 NTD Vector Production and Genetic Stability

The initial NTD particles were generated through the procedure of a virus rescue. These initial particles were amplified using a helper cell line that provided the full-length P protein to *trans*-complement the missing component of the NTD mutant.

For this amplification we first used an already existing H29 helper cell line constitutively expressing the viral P protein (Willenbrink and Neubert 1994). Later, another helper cell line based on Vero cells was generated that could also comply with GMP requirements for vaccine production (Wiegand et al. 2007).

Table 4.3 Setup of comparative production run of three different vectors ("SeV-delta*P*," "SeV-NTD," "SeV-wt") in P protein- but no C protein-expressing helper cell line

	SeV-delta*P*	SeV-NTD	SeV-wt
C proteins	–	–	+
NTD-P	–	+	–
Full-length P	+	+	+

The *P* transgene of the helper cell was designed not to encode any of the C proteins. As already discussed, the absence of C expression could be advantageous for the current purpose because the C proteins inhibit the synthesis of mRNA and antigenome templated by the negative strand genome. The recovered NTD SeV may undergo cotranscriptional editing to insert G residues after codon 317 of *P* ORF, likely generating N-terminally truncated V and W proteins. The outcomes of this remain to be evaluated. Another concern was if the truncated P protein encoded by SeV-NTD would cause any interference with the full-length P protein within the helper cell line.

However, the actual experimental results indicated that those concerns were largely groundless. After infection of cells (MOI 3, 3 cell infectious units/cell), each production run was monitored and test samples were taken every 24 h for analysis of the viral titer and production efficiency (Fig. 4.4a, Table 4.3). SeV-wt titers were the highest, as expected; however, the titer of SeV-NTD was only half of that of SeV-wt, which was surprising because the helper cell line only produced approximately 10 % of the amount of P that is produced during SeV wild-type infection (Willenbrink and Neubert 1994). This result demonstrated that SeV-NTD can be well produced in cell culture in the absence of the C proteins and full-length V and W proteins. Furthermore, when the production of SeV-NTD was compared with that of SeV-delta*P*, it was obvious that NTD-P, present during the course of production, did not cause any obvious interference with the full-length P protein. Maximum titers of $2 \times 10e7$ was most likely high enough for SeV/NTD to be used as a vaccine vector.

To complete the qualification of the helper cell line as a production system of the NTD mutant vectors, genetic stability was investigated. If any mutation occurs easily during vector production, the system would be compromised and could not be used as a reliable and safe vaccine vector. One characteristic of paramyxovirus-based viral vectors is the relatively small size of their genomes. The wild-type genome consists of 15,384 nt whereas the NTD mutant, without additional transgenes, has 15,150 nt. Sequencing of the full genome after harvesting vector RNA and reverse transcription-polymerase chain reaction (RT-PCR) could be performed easily and enabled a close monitoring of any occurring changes. This factor can contribute to the safety profile of potential future medicinal applications and represents an advantage over larger or more complex vector systems, such as adeno-viral vectors or MVA. With the here presented NTD mutant, potential mutations can only occur during the production of the vector. Because of its replication-deficient nature, further mutations in target cells or in the vaccinee are not expected. The fact that

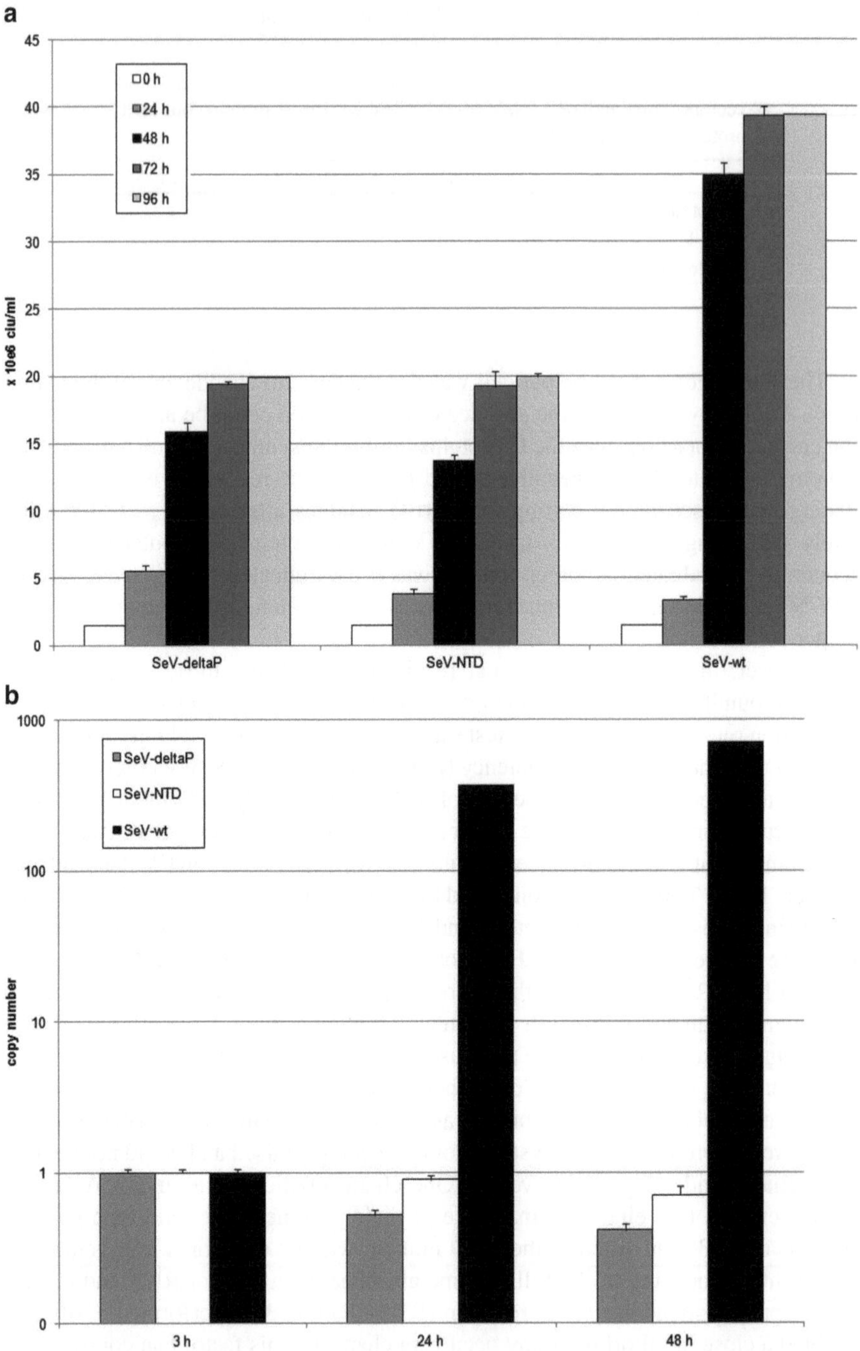

Fig. 4.4 (**a**) Production efficiency of SeV-delta*P*, SeV-NTD, and SeV-wt in a helper cell line expressing the Sendai virus P protein. (**b**) Analysis of genome replication of the viral vector SeV-NTD in comparison to vectors SeV-delta*P* and SeV-wt in non-helper cells. RNA extracts from infected cell cultures were prepared at 3, 24, and 48 h p.i. Real-time RT-PCR was performed using a Sendai vector genome-specific primer for the RT reaction first; then, a part of the *N* gene was amplified

paramyxoviruses consist of single-stranded RNA genomes and all steps of viral replication occur only in the cytoplasm without any intermediate step in the nucleus make any uptake or exchange of genomic sequences impossible according to current knowledge. To prove sequence stability, different series of GFP transgene-containing test vectors and a prototype vaccine construct (Fig. 4.1b) (both constructs are based on the NTD mutant) were analyzed for as many as ten consecutive passages in cell culture. No mutation of the vector genomes, including the NTD part of the *P* gene, was observed, except for single sporadic exchanges that did not result in changes of the amino acid sequence of the respective protein. In summary, this NTD mutant appeared to show a high level of genetic stability, and the NTD mutation could not revert to the wild-type situation.

4.4.4.2 Replication Deficiency *In Vivo* of the NTD Vectors

Replication deficiency was regarded as the ultimate prerequisite for the development of safe vaccines based on this novel vector system. To prove that no replication occurs not only in cells in culture but also *in vivo*, different setups were used. First, experiments were performed in SeV-NTD-infecting non-helper cells versus helper cells. The infections were monitored microscopically for a period of 7 days, and no viral spread of GFP expression was observed. To confirm this result by another more sensitive assay, real-time RT-PCR was applied to detect genomic viral vector templates inside infected cells at 24 and 48 h post infection. Although the wt genomic templates showed a sharp increase with time, the SeV-NTD genomic templates remained at its input level (Fig. 4.4b). The same result was obtained using a control vector with the full *P* ORF deleted ("SeV-delta*P*") (Fig. 4.4b). Thus, replication deficiency of SeV-NTD was unequivocally established.

To further substantiate this line of studies on replication deficiency of SeV-NTD, we compared the behavior of the replication-deficient vaccine rdVAC with that of replication-competent rcVAC (see Fig. 4.1b) containing the full-length *P* ORF in mice. Three days after intranasal immunization of BALB/c mice with rdVAC (2 × 10e5 CIU/mouse), tissue samples from the lung and blood were examined for the presence of infectious viral particles via microfocus assay. No viral particles could be detected in the lung of rdVAC-infected mice. In contrast, the replication-competent rcVAC was amplified there, up to 32,000 CIU/lung. In blood samples, no infective viral particles from either vector were found.

In summary, these results unequivocally demonstrated that the NTD mutant is indeed replication-deficient *in vitro* and *in vivo*.

4.4.4.3 Gene Expression Analysis

Next, we analyzed the gene expression capacity of the NTD mutant to assess its potential to stimulate sufficient immune responses as a vaccine vector. Another important factor for a potential future vaccine vector is the duration of gene/antigen

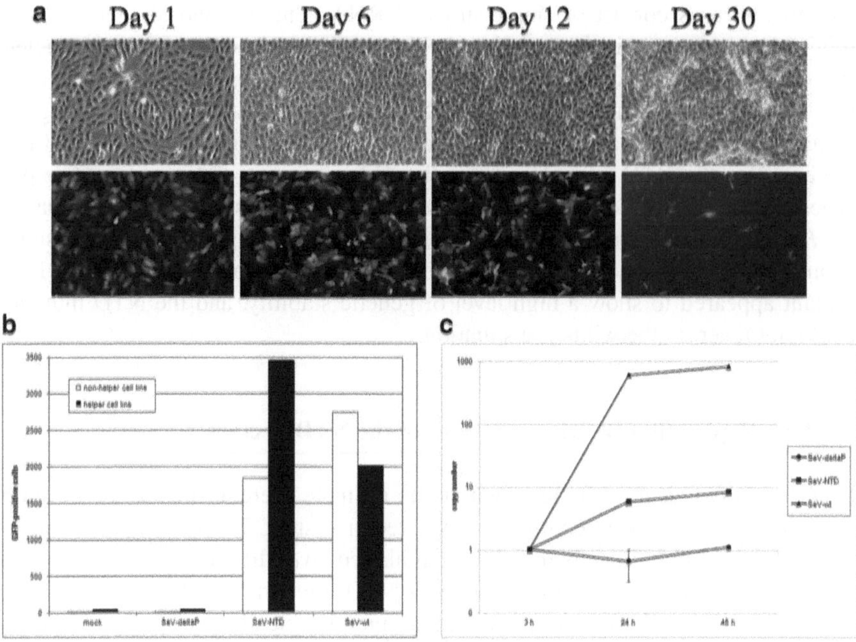

Fig. 4.5 (**a**) Monitoring of GFP expression in cell culture after infection of non-helper cells with SeV-NTD (MOI 1.0) encoding a *GFP* transgene for 30 days. (**b**) *GFP* transgene expression analysis by flow cytometry at 48 h after infection of helper and non-helper cells with SeV-NTD, the replication-deficient SeV-delta*P*, and the replication-competent SeV-wt. (**c**) Gene expression analysis of Sendai virus *N* gene mRNA via real-time RT-PCR until 48 h after infection of non-helper cells with SeV-NTD, the replication-deficient SeV-delta*P*, and the replication-competent SeV-wt

expression, because when the vector is rapidly degraded or eliminated by host response after uptake, the gene expression level might remain very basal and, thus, cannot stimulate any potent reaction. First, we analyzed GFP transgene expression of the SeV-NTD mutant in cell culture using non-helper cells. After infection at a high MOI (1 CIU/cell), GFP expression was monitored microscopically (Fig. 4.5a). Interestingly, the green fluorescence, already visible after 1 day, hardly decreased until day 6; at day 12, there were still a significant number of green fluorescent cells. This result was surprising because of previously reported nucleocapsid stability of only 24 h (Mottet et al. 1990). In addition, this result clearly indicated that the gene expression would reach significant levels and be sustained.

Efficient immune responses are considered to be dependent on a certain threshold of stimulating antigens expressed. We further quantified GFP-expressing cells in cell culture after infection at a lower MOI (0.1 CIU/cell) to exclude any saturation of gene expression through infection of cells with multiple nucleocapsids, which can occur at a higher MOI. As shown in Fig. 4.5b, the number of GFP-expressing cells in SeV-NTD infection appeared to be well above a relevant threshold not only in helper cells but also in non-helper cells. Note that the level was nearly as high as in SeV-wt infection, whereas the expression from SeV-delta*P* was virtually zero. This finding strongly suggested that SeV-NTD would fulfill the *in vivo* requirement as an antigen expression vector.

The viral *N*-gene expression in infected cells was further compared between SeV-NTD, SeV delta*P*, and the SeV-wt by real-time RT-PCR. During the course of 48 h the mRNA level in SeV-NTD-infected cells was increasing up to tenfold compared to SeV-delta*P* (Fig. 4.5c). SeV-wt expression increased steeply until 24 h. This steep increase could obviously be the result of active replication leading to the accumulation of a large amount of additional genomic templates to allow full-blown secondary transcription, which makes it difficult to assess the output of transcription of the replication-competent vector from only one genomic template. Unfortunately, single-round infection during the real *in vivo* infection situation with the replication-competent vector cannot be set up experimentally. In conclusion, the SeV vector based on the NTD mutant possesses the potential of reaching appreciable gene expression levels over a reasonable time period.

4.4.4.4 Host Responses

Then, we evaluated the *in vivo* immunogenic capacity of SeV NTD vector to depict the immunological profile of the replication-deficient vaccine vector (Wiegand et al. 2013). The replication-deficient vaccine prototype expresses the RSV F protein in a soluble form and the F and HN proteins of PIV3 as the structural components of the vector virus envelope in place of the respective proteins of the homologous SeV (see rdVAC in Fig. 4.1b). We thus included as a control the rdVAC stock that had been UV-inactivated and no longer underwent *de novo* protein synthesis. Another control included was a replication-competent counterpart, rcVAC, with the intact *P* gene (see Fig. 4.1b) to assess the transgene expression and immunogenicity under the conditions of full replication and secondary transcription. The replication-deficient vaccine prototype rdVAC was confirmed to be able to express both the RSV and the PIV3 transgenes in cell culture (Fig. 4.6a).

We wished to investigate different parameters indicating the specific humoral and cellular immune responses. As already described, SeV, as a natural respiratory pathogen, is predestined by nature to efficiently infect mucosal surfaces. For our vaccine prototype we wanted to make use of this feature and see if local mucosal immune responses could also be elicited, as this would represent a clear advantage when developing vaccines against respiratory pathogens that enter the organism via this very route. Mucosal immune responses can represent a strong first line of defense against the intruders.

BALB/c 6- to 8-week-old mice were immunized intranasally with rdVAC, UV-rdVAC (both at 1.5×10e7 CIU), or rcVAC (at 1.0×10e6 CIU). A total of three shots were applied 3 weeks apart. Ten days after the last immunization, the mice were killed and the different samples were taken for analysis: blood, nasal washes (NW), bronchoalveolar lavage (BAL), and spleen.

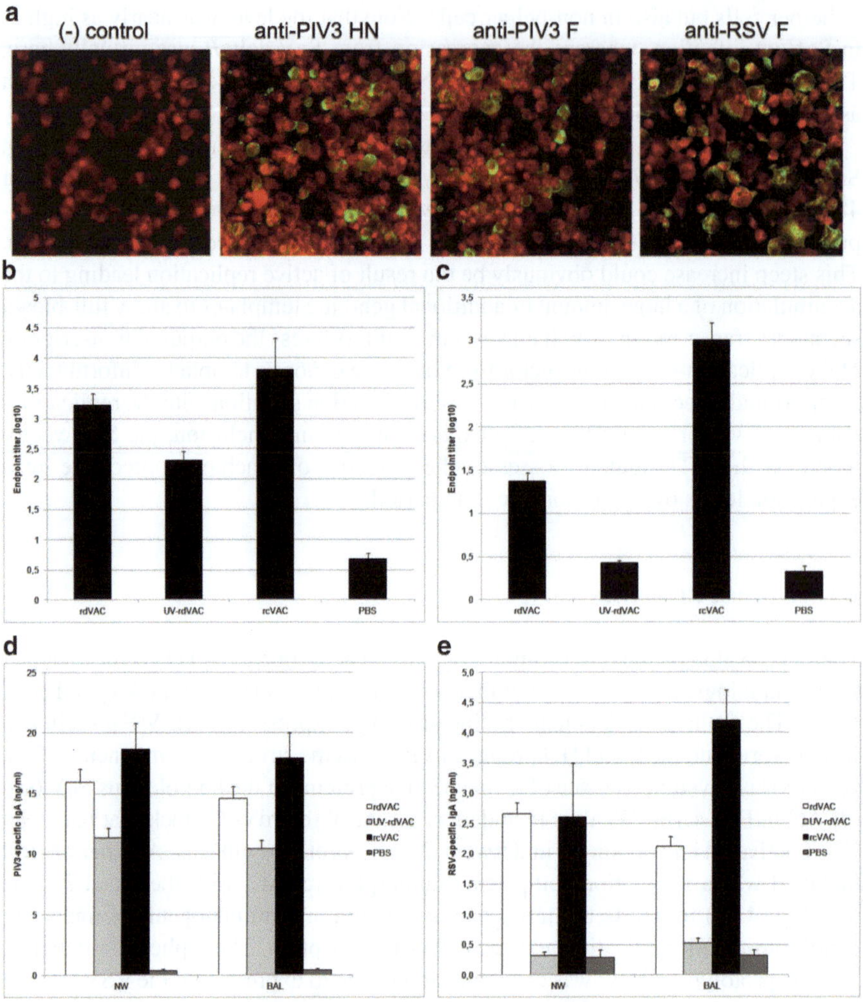

Fig. 4.6 (**a**) Gene expression analysis of PIV3 and RSV transgenes in cell culture, detected via immunofluorescence. Specific IgG titers in sera from immunized mice against PIV3 (**b**) and RSV (**c**). Comparison of serum from animals immunized with the replication-deficient vaccine (rdVAC) or the vaccine inactivated by UV light (UV-rdVAC) or the replication-competent vaccine control (rcVAC) or mock immunized (PBS). Specific IgA titer in nasal washes (NW) and bronchoalveolar lavages (BAL) from mice immunized against PIV3 (**d**) and RSV (**e**). Comparison of NW and BAL from animals immunized with the replication-deficient vaccine (rdVAC) or the vaccine inactivated by UV light (UV-rdVAC) or the replication-competent vaccine control (rcVAC) or mock immunized (PBS)

Serum Antibodies

The serum IgG titers against PIV3 were higher for rdVAC than for UV-rdVAC (Fig. 4.6b). This difference was clearly caused by whether the vector was live or killed. The difference was however, not very prominent, likely because the inactivated rdVAC still presented these antigens as structural parts of the vector particle, and the expressed PIV3 genes from the intact rdVAC were likely not high, probably because they were positioned far downstream of the 3′-end (see Fig. 4.1b). Interestingly, however, the replication-competent control elicited only a marginally higher response than did rdVAC, which might be the result of the slightly lower dose of the rcVAC, although active viral replication should easily counterbalance this lower initial input. Importantly, among serum antibodies a significant fraction of neutralizing antibodies was also found (Wiegand et al. 2013).

For the RSV F antigen the difference between rdVAC and UV-rdVAC was more evident (Fig. 4.6c). The F antigen was expressed in a soluble form and was not a structural component of the vector virus. Therefore, it was reasonable that UV-rdVAC did not present any IgG response as did the PBS (phosphate-buffered saline) control.

Mucosal Response

Remarkably, significant RSV- and PIV3-specific IgA titers could be detected in NW and in BAL (Fig. 4.6d, e), indicating that the rdVAC was also able to stimulate local mucosal response. The pattern observed for the anti-PIV3 response was similar to the serum IgG response (cf. Fig. 4.6b) in that the UV-rdVAC was still able to elicit a solid IgG or IgA response and that the levels were comparable between rdVAC and rcVAC (Fig. 4.6b, c).

The IgA responses against RSV reflect a similar pattern as the IgG response. Only rdVAC but not UV-rdVAC, as expected, was able to stimulate an antibody generation (Fig. 4.6e). Of note was that antibody titers in NW and BAL for RSV and PIV3 were present at roughly equal levels, reflecting well the common nature of respiratory tract-targeting viruses. In some other instances an imbalanced level of immunity was observed over the respiratory tissue, which could be associated with a less comprehensive state of protection (Kitikoon et al. 2006; Cutter et al. 2002).

Cellular Responses

For vaccine development against some pathogens against which no vaccine yet exists, such as RSV, an effective immunization might have to be based on different arms of the immune system, that is, the cellular response. As shown earlier, a replication-competent SeV vaccine vector was able to confer protection against RSV, even in antibody-deficient transgenic mice, so long as an efficient T-cell response was elicited (Voges et al. 2007). The importance of CTL for protection also has been demonstrated for simian immunodeficiency virus (SIV) (see Chap. 5). We therefore analyzed the capacity of rdVAC, in comparison with UV-rdVAC, to stimulate a CTL response. Splenocytes isolated from immunized animals were incubated with P815 target cells that were infected with either PIV3 or RSV. PIV3-infected target cells were lysed efficiently through lymphocytes isolated from animals immunized not only with rdVAC but also with UV-rdVAC (Fig. 4.7a). The vector particles

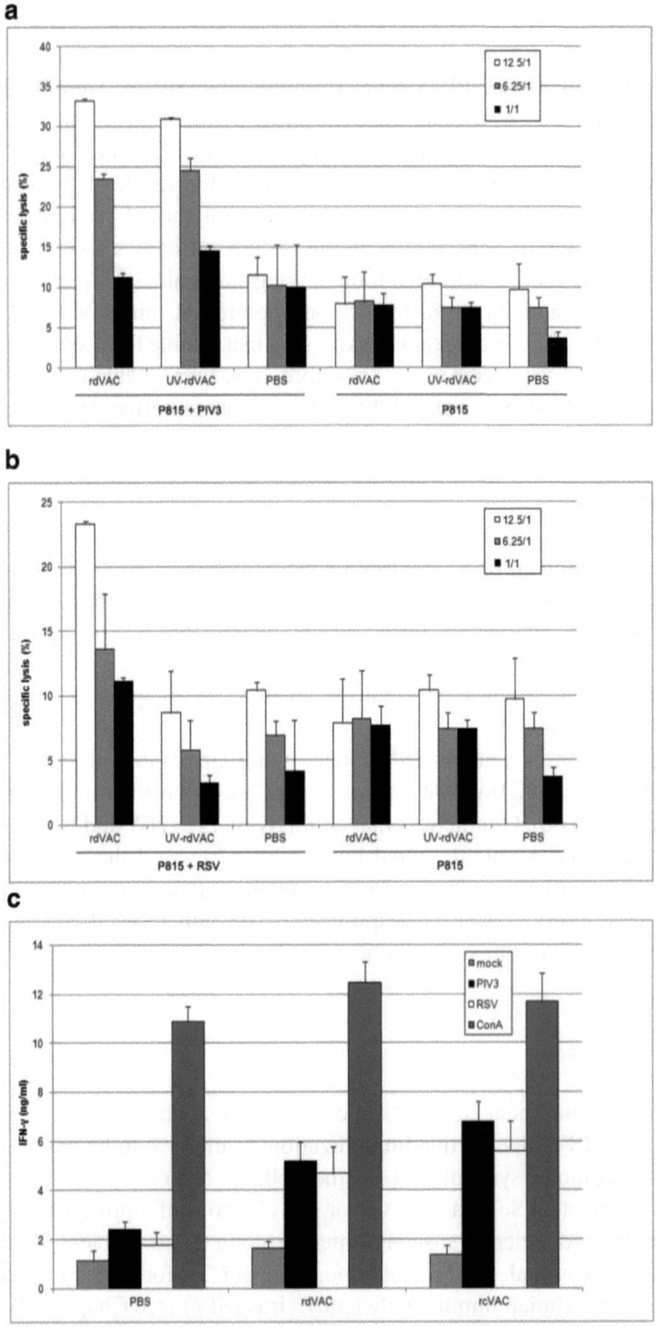

Fig. 4.7 Analysis of cytotoxic T-lymphocyte (CTL) response against (**a**) PIV3 and (**b**) RSV. Splenocytes restimulated with either PIV3 or RSV particles were incubated at the ratios 12.5/1, 6.25/1, and 1/1 with P815 target cells that were previously infected with PIV3 or RSV and radioactively labeled. Specific lysis was detected through release of radioactively labeled metabolites. Uninfected P815 cells were used as control. (**c**) IFN-γ expression from restimulated splenocytes. Splenocytes were isolated from immunized animals and restimulated with either PIV3 or RSV particles. Animals were immunized with rdVAC or rcVAC

themselves thus appeared to be able to confer T-cell response. The impact of active gene expression could not be clearly assessed from these data. However, when the response against RSV was analyzed, the observed response can only be caused by actively expressed antigen. As shown in Fig. 4.7b, only rdVAC immunized mice were able to generate an efficient CTL response, whereas UV-rdVAC could not elicit any response.

After we demonstrated that the replication-deficient vector is able to stimulate a substantial cytotoxic T-cell response, we were interested in analyzing the T-helper (Th) cell response in more detail. T-cell subsets are known to affect and support different functions during an immune response and, therefore, we performed a cytokine analysis to gain insight into Th subsets, Th1/Th2, that might be stimulated through vaccination with the replication-deficient vaccine vector. Analysis of IL-5 as a typical representative for a Th2 response, mainly associated with supporting humoral responses, could not be detected from restimulated splenocytes. On the other hand, secretion of IFN-γ associated with Th1 response could very well be detected (Fig. 4.7c).

Overall, our results showed that the replication-deficient prototype vaccine could elicit a broad spectrum of immune responses against RSV and PIV3, indicating that it would represent an interesting vaccine platform for the development of safe vaccines in general, particularly for the respiratory tract-targeting viruses.

4.5 Conclusions

SeV qualifies as a vector system for safe and efficient applications based on several advantageous factors intrinsic to this virus: SeV is a nonhuman pathogen, there is no pre-immunity against SeV in humans, SeV contains a single-stranded RNA genome that replicates entirely in the cytoplasm without any DNA intermediates, SeV shows strong tissue restriction, and SeV can be applied mucosally. In addition, a replication-deficient vector contributes additional factors that further increase safety: no uncontrolled spread in the organism, no persistence of the viral vector, no reversion to the wild-type, and little possibility of any other unforeseen mutation.

The aim of the research presented here was to obtain a 100 % replication-incompetent SeV vector that was still able to perform efficient gene expression.

Replication deficiency was accomplished through modification of the replication complex that consists of the viral N, P, and L proteins. However, neither N nor L could be modified in a way that enabled full replication deficiency while still performing relevant gene expression. Finally, uncoupling of transcription from genome replication could be accomplished through specific engineering of the polymerase cofactor P protein. The resulting SeV vector appeared to meet all necessary characteristics:

- Production in cell culture up to good titers.
- Replication deficiency as proven in cell culture and *in vivo*.
- Genetic stability.

- Good gene expression performance, lasting over several days.
- Ability to stimulate broad specific immune responses consisting of systemic humoral, local mucosal, T-helper, and CTL responses.

After achieving this preclinical proof-of-concept, we consider this vector system shall be qualified for specific vaccine developments. Based on the herein-described characteristics of the replication-deficient SeV vector, this platform could be used for the development of vaccines with an enhanced safety profile. Safety is a most important issue in the development of vaccines, especially for vaccines that target specific populations at risk. With vaccines already existing on the market, children, infants, elderly, or immunocompromised people very often are vaccinated. Furthermore, there is still a lack of vaccines against several pathogens despite a high unmet medical need. So far, using the classical vaccine approaches, vaccines against RSV, HIV, HCV, dengue, or malaria—to name just a few—have not been developed. Target populations for these diseases include certain risk groups. We hope that our novel replication-deficient vector system will be able to contribute to the development of safe and efficient vaccines to help preventing diseases in the future. Because of its efficient applicability via the mucosal route, this SeV vector is especially well suited for vaccine developments against respiratory pathogens, such as RSV, PIV3, or influenza.

Besides the traditional prophylactic vaccination strategy, this SeV vector system can also be considered useful for various other applications where safe gene delivery tools are needed. Therapeutic vaccination against infectious diseases or cancer, programming or reprogramming of cells or, simply, use as a research tool as a non-cytotoxic gene expression vector, could be potential further applications of this replication-deficient SeV vector platform.

References

Alwan WH, Record FM, Openshaw PJ (1992) CD4+ T cells clear virus but augment disease in mice infected with respiratory syncytial virus. Comparison with the effects of CD8+ T cells. Clin Exp Immunol 88(3):527–536

Anderson LJ, Hierholzer JC, Tsou C, Hendry RM, Fernie BF, Stone Y, McIntosh K (1985) Antigenic characterization of respiratory syncytial virus strains with monoclonal antibodies. J Infect Dis 151(4):626–633

Bossow S, Schlecht S, Schubbert R, Pfeiffer M, Neubert WJ, Wiegand M (2012) Evaluation of nucleocapsid and phosphoprotein P functionality as critical factors during the early phase of paramyxoviral infection. Open Virol J 6:73–81. doi:10.2174/1874357901206010073

Boxus M, Tignon M, Roels S, Toussaint JF, Walravens K, Benoit MA, Coppe P, Letesson JJ, Letellier C, Kerkhofs P (2007) DNA immunization with plasmids encoding fusion and nucleocapsid proteins of bovine respiratory syncytial virus induces a strong cell-mediated immunity and protects calves against challenge. J Virol 81(13):6879–6889. doi:10.1128/jvi.00502-07

Buchholz CJ, Spehner D, Drillien R, Neubert WJ, Homann HE (1993) The conserved N-terminal region of Sendai virus nucleocapsid protein NP is required for nucleocapsid assembly. J Virol 67(10):5803–5812

Bukreyev A, Yang L, Fricke J, Cheng L, Ward JM, Murphy BR, Collins PL (2008) The secreted form of respiratory syncytial virus G glycoprotein helps the virus evade antibody-mediated restriction of replication by acting as an antigen decoy and through effects on Fc receptor-bearing leukocytes. J Virol 82(24):12191–12204. doi:10.1128/jvi.01604-08

Cadd T, Garcin D, Tapparel C, Itoh M, Homma M, Roux L, Curran J, Kolakofsky D (1996) The Sendai paramyxovirus accessory C proteins inhibit viral genome amplification in a promoter-specific fashion. J Virol 70(8):5067–5074

Cardenas S, Auais A, Piedimonte G (2005) Palivizumab in the prophylaxis of respiratory syncytial virus infection. Expert Rev Anti Infect Ther 3(5):719–726. doi:10.1586/14787210.3.5.719

Cherrie AH, Anderson K, Wertz GW, Openshaw PJ (1992) Human cytotoxic T cells stimulated by antigen on dendritic cells recognize the N, SH, F, M, 22K, and 1b proteins of respiratory syncytial virus. J Virol 66(4):2102–2110

Collins PL, Crowe JE Jr (2007) Respiratory syncytial virus and metapneumovirus. In: Knipe DM, Howley P (eds) Fields virology, 5th edn. Lippincott Williams & Wilkins, Philadelphia, pp 1601–1646

Connors M, Giese NA, Kulkarni AB, Firestone CY, Morse HC 3rd, Murphy BR (1994) Enhanced pulmonary histopathology induced by respiratory syncytial virus (RSV) challenge of formalin-inactivated RSV-immunized BALB/c mice is abrogated by depletion of interleukin-4 (IL-4) and IL-10. J Virol 68(8):5321–5325

Curran J, Boeck R, Kolakofsky D (1991) The Sendai virus P gene expresses both an essential protein and an inhibitor of RNA synthesis by shuffling modules via mRNA editing. EMBO J 10(10):3079–3085

Curran J, Marq JB, Kolakofsky D (1992) The Sendai virus nonstructural C proteins specifically inhibit viral mRNA synthesis. Virology 189(2):647–656

Curran J, Homann H, Buchholz C, Rochat S, Neubert W, Kolakofsky D (1993) The hypervariable C-terminal tail of the Sendai paramyxovirus nucleocapsid protein is required for template function but not for RNA encapsidation. J Virol 67(7):4358–4364

Curran J, Pelet T, Kolakofsky D (1994) An acidic activation-like domain of the Sendai virus P protein is required for RNA synthesis and encapsidation. Virology 202(2):875–884. doi:10.1006/viro.1994.1409

Curran J, Boeck R, Lin-Marq N, Lupas A, Kolakofsky D (1995a) Paramyxovirus phosphoproteins form homotrimers as determined by an epitope dilution assay, via predicted coiled coils. Virology 214(1):139–149. doi:10.1006/viro.1995.9946

Curran J, Marq JB, Kolakofsky D (1995b) An N-terminal domain of the Sendai paramyxovirus P protein acts as a chaperone for the NP protein during the nascent chain assembly step of genome replication. J Virol 69(2):849–855

Cutter D, Mason KW, Howell AP, Fink DL, Green BA, St. Geme JW 3rd (2002) Immunization with *Haemophilus influenzae* Hap adhesin protects against nasopharyngeal colonization in experimental mice. J Infect Dis 186(8):1115–1121. doi:10.1086/344233

de Swart RL, Kuiken T, Timmerman HH, van Amerongen G, Van Den Hoogen BG, Vos HW, Neijens HJ, Andeweg AC, Osterhaus AD (2002) Immunization of macaques with formalin-inactivated respiratory syncytial virus (RSV) induces interleukin-13-associated hypersensitivity to subsequent RSV infection. J Virol 76(22):11561–11569

de Waal L, Wyatt LS, Yuksel S, van Amerongen G, Moss B, Niesters HG, Osterhaus AD, de Swart RL (2004) Vaccination of infant macaques with a recombinant modified vaccinia virus Ankara expressing the respiratory syncytial virus F and G genes does not predispose for immunopathology. Vaccine 22(8):923–926. doi:10.1016/j.vaccine.2003.10.010

Einberger H, Mertz R, Hofschneider PH, Neubert WJ (1990) Purification, renaturation, and reconstituted protein kinase activity of the Sendai virus large (L) protein: L protein phosphorylates the NP and P proteins in vitro. J Virol 64(9):4274–4280

Falsey AR, Walsh EE (1996) Safety and immunogenicity of a respiratory syncytial virus subunit vaccine (PFP-2) in ambulatory adults over age 60. Vaccine 14(13):1214–1218

Falsey AR, Hennessey PA, Formica MA, Cox C, Walsh EE (2005) Respiratory syncytial virus infection in elderly and high-risk adults. N Engl J Med 352(17):1749–1759. doi:10.1056/NEJMoa043951

Feller JA, Smallwood S, Horikami SM, Moyer SA (2000) Mutations in conserved domains IV and VI of the large (L) subunit of the Sendai virus RNA polymerase give a spectrum of defective RNA synthesis phenotypes. Virology 269(2):426–439. doi:10.1006/viro.2000.0234

Fu Y, He J, Zheng X, Wu Q, Zhang M, Wang X, Wang Y, Xie C, Tang Q, Wei W, Wang M, Song J, Qu J, Zhang Y, Wang X, Hong T (2009) Intranasal immunization with a replication-deficient adenoviral vector expressing the fusion glycoprotein of respiratory syncytial virus elicits protective immunity in BALB/c mice. Biochem Biophys Res Commun 381(4):528–532. doi:10.1016/j.bbrc.2009.02.075

Garcin D, Latorre P, Kolakofsky D (1999) Sendai virus C proteins counteract the interferon-mediated induction of an antiviral state. J Virol 73(8):6559–6565

Geall AJ, Verma A, Otten GR, Shaw CA, Hekele A, Banerjee K, Cu Y, Beard CW, Brito LA, Krucker T, O'Hagan DT, Singh M, Mason PW, Valiante NM, Dormitzer PR, Barnett SW, Rappuoli R, Ulmer JB, Mandl CW (2012) Nonviral delivery of self-amplifying RNA vaccines. Proc Natl Acad Sci USA 109(36):14604–14609. doi:10.1073/pnas.1209367109

Glenn GM, Smith G, Fries L, Raghunandan R, Lu H, Zhou B, Thomas DN, Hickman SP, Kpamegan E, Boddapati S, Piedra PA (2013) Safety and immunogenicity of a Sf9 insect cell-derived respiratory syncytial virus fusion protein nanoparticle vaccine. Vaccine 31(3):524–532. doi:10.1016/j.vaccine.2012.11.009

Glezen WP, Taber LH, Frank AL, Kasel JA (1986) Risk of primary infection and reinfection with respiratory syncytial virus. Am J Dis Child 140(6):543–546 (1960)

Gomez M, Mufson MA, Dubovsky F, Knightly C, Zeng W, Losonsky G (2009) Phase-I study MEDI-534, of a live, attenuated intranasal vaccine against respiratory syncytial virus and parainfluenza-3 virus in seropositive children. Pediatr Infect Dis J 28(7):655–658. doi:10.1097/INF.0b013e318199c3b1

Gotoh B, Takeuchi K, Komatsu T, Yokoo J, Kimura Y, Kurotani A, Kato A, Nagai Y (1999) Knockout of the Sendai virus C gene eliminates the viral ability to prevent the interferon-alpha/beta-mediated responses. FEBS Lett 459(2):205–210

Griesenbach U, Cassady RL, Ferrari S, Fukumura M, Muller C, Schmitt E, Zhu J, Jeffery PK, Nagai Y, Geddes DM, Hasegawa M, Alton EW (2002) The nasal epithelium as a factory for systemic protein delivery. Mol Ther 5(2):98–103. doi:10.1006/mthe.2002.0524

The IMpact-RSV Study Group (1998) Palivizumab, a humanized respiratory syncytial virus monoclonal antibody, reduces hospitalization from respiratory syncytial virus infection in high-risk infants. Pediatrics 102(3 Pt 1):531–537

Hammond DC, Lesnaw JA (1987) Functional analysis of hypomethylation variants of the New Jersey serotype of vesicular stomatitis virus. Virology 160(2):330–335

Hammond DC, Haley BE, Lesnaw JA (1992) Identification and characterization of serine/threonine protein kinase activity intrinsic to the L protein of vesicular stomatitis virus New Jersey. J Gen Virol 73(Pt 1):67–75

Hasan MK, Kato A, Muranaka M, Yamaguchi R, Sakai Y, Hatano I, Tashiro M, Nagai Y (2000) Versatility of the accessory C proteins of Sendai virus: contribution to virus assembly as an additional role. J Virol 74(12):5619–5628

Heidema J, de Bree GJ, De Graaff PM, van Maren WW, Hoogerhout P, Out TA, Kimpen JL, van Bleek GM (2004) Human CD8(+) T cell responses against five newly identified respiratory syncytial virus-derived epitopes. J Gen Virol 85(Pt 8):2365–2374. doi:10.1099/vir.0.80131-0

Hercyk N, Horikami SM, Moyer SA (1988) The vesicular stomatitis virus L protein possesses the mRNA methyltransferase activities. Virology 163(1):222–225

Horikami SM, Curran J, Kolakofsky D, Moyer SA (1992) Complexes of Sendai virus NP-P and P-L proteins are required for defective interfering particle genome replication in vitro. J Virol 66(8):4901–4908

Horikami SM, Smallwood S, Moyer SA (1996) The Sendai virus V protein interacts with the NP protein to regulate viral genome RNA replication. Virology 222(2):383–390. doi:10.1006/viro.1996.0435

Horikami SM, Hector RE, Smallwood S, Moyer SA (1997) The Sendai virus C protein binds the L polymerase protein to inhibit viral RNA synthesis. Virology 235(2):261–270. doi:10.1006/viro.1997.8702

Hunt DM, Hutchinson KL (1993) Amino acid changes in the L polymerase protein of vesicular stomatitis virus which confer aberrant polyadenylation and temperature-sensitive phenotypes. Virology 193(2):786–793. doi:10.1006/viro.1993.1187

Irie T, Shimazu Y, Yoshida T, Sakaguchi T (2007) The YLDL sequence within Sendai virus M protein is critical for budding of virus-like particles and interacts with Alix/AIP1 independently of C protein. J Virol 81(5):2263–2273. doi:10.1128/jvi.02218-06

Jin H, Elliott RM (1992) Mutagenesis of the L protein encoded by Bunyamwera virus and production of monospecific antibodies. J Gen Virol 73(Pt 9):2235–2244

Jin H, Zhou H, Cheng X, Tang R, Munoz M, Nguyen N (2000) Recombinant respiratory syncytial viruses with deletions in the NS1, NS2, SH, and M2-2 genes are attenuated in vitro and in vivo. Virology 273(1):210–218. doi:10.1006/viro.2000.0393

Johnson TR, Graham BS (1999) Secreted respiratory syncytial virus G glycoprotein induces interleukin-5 (IL-5), IL-13, and eosinophilia by an IL-4-independent mechanism. J Virol 73(10):8485–8495

Kapikian AZ, Mitchell RH, Chanock RM, Shvedoff RA, Stewart CE (1969) An epidemiologic study of altered clinical reactivity to respiratory syncytial (RS) virus infection in children previously vaccinated with an inactivated RS virus vaccine. Am J Epidemiol 89(4):405–421

Karron RA, Wright PF, Crowe JE Jr, Clements-Mann ML, Thompson J, Makhene M, Casey R, Murphy BR (1997) Evaluation of two live, cold-passaged, temperature-sensitive respiratory syncytial virus vaccines in chimpanzees and in human adults, infants, and children. J Infect Dis 176(6):1428–1436

Karron RA, Wright PF, Belshe RB, Thumar B, Casey R, Newman F, Polack FP, Randolph VB, Deatly A, Hackell J, Gruber W, Murphy BR, Collins PL (2005) Identification of a recombinant live attenuated respiratory syncytial virus vaccine candidate that is highly attenuated in infants. J Infect Dis 191(7):1093–1104. doi:10.1086/427813

Kato A, Kiyotani K, Sakai Y, Yoshida T, Nagai Y (1997a) The paramyxovirus, Sendai virus, V protein encodes a luxury function required for viral pathogenesis. EMBO J 16(3):578–587. doi:10.1093/emboj/16.3.578

Kato A, Kiyotani K, Sakai Y, Yoshida T, Shioda T, Nagai Y (1997b) Importance of the cysteine-rich carboxyl-terminal half of V protein for Sendai virus pathogenesis. J Virol 71(10):7266–7272

Kitikoon P, Nilubol D, Erickson BJ, Janke BH, Hoover TC, Sornsen SA, Thacker EL (2006) The immune response and maternal antibody interference to a heterologous H1N1 swine influenza virus infection following vaccination. Vet Immunol Immunopathol 112(3–4):117–128. doi:10.1016/j.vetimm.2006.02.008

Lamb RA, Mahy BW, Choppin PW (1976) The synthesis of Sendai virus polypeptides in infected cells. Virology 69(1):116–131

Langley JM, Sales V, McGeer A, Guasparini R, Predy G, Meekison W, Li M, Capellan J, Wang E (2009) A dose-ranging study of a subunit respiratory syncytial virus subtype A vaccine with and without aluminum phosphate adjuvantation in adults > or =65 years of age. Vaccine 27(42):5913–5919. doi:10.1016/j.vaccine.2009.07.038

Latorre P, Kolakofsky D, Curran J (1998) Sendai virus Y proteins are initiated by a ribosomal shunt. Mol Cell Biol 18(9):5021–5031

Levely ME, Bannow CA, Smith CW, Nicholas JA (1991) Immunodominant T-cell epitope on the F protein of respiratory syncytial virus recognized by human lymphocytes. J Virol 65(7):3789–3796

Mok H, Lee S, Utley TJ, Shepherd BE, Polosukhin VV, Collier ML, Davis NL, Johnston RE, Crowe JE Jr (2007) Venezuelan equine encephalitis virus replicon particles encoding respiratory syncytial virus surface glycoproteins induce protective mucosal responses in mice and cotton rats. J Virol 81(24):13710–13722. doi:10.1128/jvi.01351-07

Moore EC, Barber J, Tripp RA (2008) Respiratory syncytial virus (RSV) attachment and nonstructural proteins modify the type I interferon response associated with suppressor of cytokine signaling (SOCS) proteins and IFN-stimulated gene-15 (ISG15). Virol J 5:116. doi:10.1186/1743-422x-5-116

Moriya C, Horiba S, Inoue M, Iida A, Hara H, Shu T, Hasegawa M, Matano T (2008) Antigen-specific T-cell induction by vaccination with a recombinant Sendai virus vector even in the

presence of vector-specific neutralizing antibodies in rhesus macaques. Biochem Biophys Res Commun 371(4):850–854. doi:10.1016/j.bbrc.2008.04.156
Moriya C, Horiba S, Kurihara K, Kamada T, Takahara Y, Inoue M, Iida A, Hara H, Shu T, Hasegawa M, Matano T (2011) Intranasal Sendai viral vector vaccination is more immunogenic than intramuscular under pre-existing anti-vector antibodies. Vaccine 29(47):8557–8563. doi:10.1016/j.vaccine.2011.09.028
Mottet G, Curran J, Roux L (1990) Intracellular stability of nonreplicating paramyxovirus nucleocapsids. Virology 176(1):1–7
Murawski MR, McGinnes LW, Finberg RW, Kurt-Jones EA, Massare MJ, Smith G, Heaton PM, Fraire AE, Morrison TG (2010) Newcastle disease virus-like particles containing respiratory syncytial virus G protein induced protection in BALB/c mice, with no evidence of immunopathology. J Virol 84(2):1110–1123. doi:10.1128/jvi.01709-09
Nair H, Nokes DJ, Gessner BD, Dherani M, Madhi SA, Singleton RJ, O'Brien KL, Roca A, Wright PF, Bruce N, Chandran A, Theodoratou E, Sutanto A, Sedyaningsih ER, Ngama M, Munywoki PK, Kartasasmita C, Simoes EA, Rudan I, Weber MW, Campbell H (2010) Global burden of acute lower respiratory infections due to respiratory syncytial virus in young children: a systematic review and meta-analysis. Lancet 375(9725):1545–1555. doi:10.1016/s0140-6736(10)60206-1
Olszewska W, Suezer Y, Sutter G, Openshaw PJ (2004) Protective and disease-enhancing immune responses induced by recombinant modified vaccinia Ankara (MVA) expressing respiratory syncytial virus proteins. Vaccine 23(2):215–221. doi:10.1016/j.vaccine.2004.05.015
Openshaw PJ, Culley FJ, Olszewska W (2001) Immunopathogenesis of vaccine-enhanced RSV disease. Vaccine 20(suppl 1):S27–S31
Patwardhan S, Gupta KC (1988) Translation initiation potential of the 5′ proximal AUGs of the polycistronic P/C mRNA of Sendai virus. A multipurpose vector for site-specific mutagenesis. J Biol Chem 263(10):4907–4913
Piedra PA, Cron SG, Jewell A, Hamblett N, McBride R, Palacio MA, Ginsberg R, Oermann CM, Hiatt PW (2003) Immunogenicity of a new purified fusion protein vaccine to respiratory syncytial virus: a multi-center trial in children with cystic fibrosis. Vaccine 21(19–20):2448–2460
Poch O, Blumberg BM, Bougueleret L, Tordo N (1990) Sequence comparison of five polymerases (L proteins) of unsegmented negative-strand RNA viruses: theoretical assignment of functional domains. J Gen Virol 71(Pt 5):1153–1162
Power UF, Nguyen TN, Rietveld E, de Swart RL, Groen J, Osterhaus AD, de Groot R, Corvaia N, Beck A, Bouveret-Le-Cam N, Bonnefoy JY (2001) Safety and immunogenicity of a novel recombinant subunit respiratory syncytial virus vaccine (BBG2Na) in healthy young adults. J Infect Dis 184(11):1456–1460. doi:10.1086/324426
Pringle CR, Filipiuk AH, Robinson BS, Watt PJ, Higgins P, Tyrrell DA (1993) Immunogenicity and pathogenicity of a triple temperature-sensitive modified respiratory syncytial virus in adult volunteers. Vaccine 11(4):473–478
Quan FS, Kim Y, Lee S, Yi H, Kang SM, Bozja J, Moore ML, Compans RW (2011) Viruslike particle vaccine induces protection against respiratory syncytial virus infection in mice. J Infect Dis 204(7):987–995. doi:10.1093/infdis/jir474
Ryan KW, Morgan EM, Portner A (1991) Two noncontiguous regions of Sendai virus P protein combine to form a single nucleocapsid binding domain. Virology 180(1):126–134
Schnell MJ, Conzelmann KK (1995) Polymerase activity of in vitro mutated rabies virus L protein. Virology 214(2):522–530. doi:10.1006/viro.1995.0063
Shay DK, Holman RC, Newman RD, Liu LL, Stout JW, Anderson LJ (1999) Bronchiolitis-associated hospitalizations among US children, 1980–1996. JAMA 282(15):1440–1446
Sidhu MS, Menonna JP, Cook SD, Dowling PC, Udem SA (1993) Canine distemper virus L gene: sequence and comparison with related viruses. Virology 193(1):50–65. doi:10.1006/viro.1993.1102
Sleat DE, Banerjee AK (1993) Transcriptional activity and mutational analysis of recombinant vesicular stomatitis virus RNA polymerase. J Virol 67(3):1334–1339
Slobod KS, Shenep JL, Lujan-Zilbermann J, Allison K, Brown B, Scroggs RA, Portner A, Coleclough C, Hurwitz JL (2004) Safety and immunogenicity of intranasal murine parainflu-

enza virus type 1 (Sendai virus) in healthy human adults. Vaccine 22(23-24):3182–3186. doi:10.1016/j.vaccine.2004.01.053

Smallwood S, Ryan KW, Moyer SA (1994) Deletion analysis defines a carboxyl-proximal region of Sendai virus P protein that binds to the polymerase L protein. Virology 202(1):154–163. doi:10.1006/viro.1994.1331

Smallwood S, Easson CD, Feller JA, Horikami SM, Moyer SA (1999) Mutations in conserved domain II of the large (L) subunit of the Sendai virus RNA polymerase abolish RNA synthesis. Virology 262(2):375–383. doi:10.1006/viro.1999.9933

Smallwood S, Cevik B, Moyer SA (2002a) Intragenic complementation and oligomerization of the L subunit of the Sendai virus RNA polymerase. Virology 304(2):235–245

Smallwood S, Hovel T, Neubert WJ, Moyer SA (2002b) Different substitutions at conserved amino acids in domains II and III in the Sendai L RNA polymerase protein inactivate viral RNA synthesis. Virology 304(1):135–145

Smith G, Raghunandan R, Wu Y, Liu Y, Massare M, Nathan M, Zhou B, Lu H, Boddapati S, Li J, Flyer D, Glenn G (2012) Respiratory syncytial virus fusion glycoprotein expressed in insect cells form protein nanoparticles that induce protective immunity in cotton rats. PLoS One 7(11):e50852. doi:10.1371/journal.pone.0050852

Srikiatkhachorn A, Braciale TJ (1997) Virus-specific CD8+ T lymphocytes downregulate T helper cell type 2 cytokine secretion and pulmonary eosinophilia during experimental murine respiratory syncytial virus infection. J Exp Med 186(3):421–432

Stegmann T, Kamphuis T, Meijerhof T, Goud E, de Haan A, Wilschut J (2010) Lipopeptide-adjuvanted respiratory syncytial virus virosomes: a safe and immunogenic non-replicating vaccine formulation. Vaccine 28(34):5543–5550. doi:10.1016/j.vaccine.2010.06.041

Tapparel C, Hausmann S, Pelet T, Curran J, Kolakofsky D, Roux L (1997) Inhibition of Sendai virus genome replication due to promoter-increased selectivity: a possible role for the accessory C proteins. J Virol 71(12):9588–9599

Ternette N, Tippler B, Uberla K, Grunwald T (2007) Immunogenicity and efficacy of codon optimized DNA vaccines encoding the F-protein of respiratory syncytial virus. Vaccine 25(41):7271–7279. doi:10.1016/j.vaccine.2007.07.025

van Bleek GM, Poelen MC, van der Most R, Brugghe HF, Timmermans HA, Boog CJ, Hoogerhout P, Otten HG, van Els CA (2003) Identification of immunodominant epitopes derived from the respiratory syncytial virus fusion protein that are recognized by human CD4 T cells. J Virol 77(2):980–988

Vidal S, Curran J, Kolakofsky D (1990) Editing of the Sendai virus P/C mRNA by G insertion occurs during mRNA synthesis via a virus-encoded activity. J Virol 64(1):239–246

Voges B, Vallbracht S, Zimmer G, Bossow S, Neubert WJ, Richter K, Hobeika E, Herrler G, Ehl S (2007) Recombinant Sendai virus induces T cell immunity against respiratory syncytial virus that is protective in the absence of antibodies. Cell Immunol 247(2):85–94. doi:10.1016/j.cellimm.2007.07.005

Waris ME, Tsou C, Erdman DD, Zaki SR, Anderson LJ (1996) Respiratory synctial virus infection in BALB/c mice previously immunized with formalin-inactivated virus induces enhanced pulmonary inflammatory response with a predominant Th2-like cytokine pattern. J Virol 70(5):2852–2860

Wiegand MA, Bossow S, Schlecht S, Neubert WJ (2007) De novo synthesis of N and P proteins as a key step in Sendai virus gene expression. J Virol 81(24):13835–13844. doi:10.1128/jvi.00914-07

Wiegand M, Gori-Savellini G, Martorelli B, Bossow S, Neubert WJ, Cusi MG (2013) Evaluation of a novel immunogenic vaccine platform based on a genome replication-deficient Sendai vector. Vaccine. doi:10.1016/j.vaccine.2013.06.053

Willenbrink W, Neubert WJ (1994) Long-term replication of Sendai virus defective interfering particle nucleocapsids in stable helper cell lines. J Virol 68(12):8413–8417

Wright PF, Karron RA, Belshe RB, Thompson J, Crowe JE Jr, Boyce TG, Halburnt LL, Reed GW, Whitehead SS, Anderson EL, Wittek AE, Casey R, Eichelberger M, Thumar B, Randolph VB, Udem SA, Chanock RM, Murphy BR (2000) Evaluation of a live, cold-passaged, temperature-sensitive, respiratory syncytial virus vaccine candidate in infancy. J Infect Dis 182(5):1331–1342. doi:10.1086/315859

Wright PF, Karron RA, Madhi SA, Treanor JJ, King JC, O'Shea A, Ikizler MR, Zhu Y, Collins PL, Cutland C, Randolph VB, Deatly AM, Hackell JG, Gruber WC, Murphy BR (2006) The interferon antagonist NS2 protein of respiratory syncytial virus is an important virulence determinant for humans. J Infect Dis 193(4):573–581. doi:10.1086/499600

Wyatt LS, Whitehead SS, Venanzi KA, Murphy BR, Moss B (1999) Priming and boosting immunity to respiratory syncytial virus by recombinant replication-defective vaccinia virus MVA. Vaccine 18(5–6):392–397

Yonemitsu Y, Kitson C, Ferrari S, Farley R, Griesenbach U, Judd D, Steel R, Scheid P, Zhu J, Jeffery PK, Kato A, Hasan MK, Nagai Y, Masaki I, Fukumura M, Hasegawa M, Geddes DM, Alton EW (2000) Efficient gene transfer to airway epithelium using recombinant Sendai virus. Nat Biotechnol 18(9):970–973. doi:10.1038/79463

Yu D, Shioda T, Kato A, Hasan MK, Sakai Y, Nagai Y (1997) Sendai virus-based expression of HIV-1 gp120: reinforcement by the V(−) version. Genes Cells 2(7):457–466

Yu JR, Kim S, Lee JB, Chang J (2008) Single intranasal immunization with recombinant adenovirus-based vaccine induces protective immunity against respiratory syncytial virus infection. J Virol 82(5):2350–2357. doi:10.1128/jvi.02372-07

Zhan X, Hurwitz JL, Krishnamurthy S, Takimoto T, Boyd K, Scroggs RA, Surman S, Portner A, Slobod KS (2007) Respiratory syncytial virus (RSV) fusion protein expressed by recombinant Sendai virus elicits B-cell and T-cell responses in cotton rats and confers protection against RSV subtypes A and B. Vaccine 25(52):8782–8793. doi:10.1016/j.vaccine.2007.10.038

Zhou H, Thompson WW, Viboud CG, Ringholz CM, Cheng PY, Steiner C, Abedi GR, Anderson LJ, Brammer L, Shay DK (2012) Hospitalizations associated with influenza and respiratory syncytial virus in the United States, 1993–2008. Clin Infect Dis 54(10):1427–1436. doi:10.1093/cid/cis211

Chapter 5
Development of Vaccines Using SeV Vectors Against AIDS and Other Infectious Diseases

Sayuri Seki and Tetsuro Matano

Abstract Development of effective vaccines, essential for the control of a variety of infectious diseases, requires optimization of antigens and delivery systems. Sendai virus (SeV) vector has been recognized as a promising, novel vaccine delivery tool. We have shown the potential of a recombinant SeV vector vaccine to efficiently induce antigen-specific cytotoxic T-lymphocyte (CTL) responses, leading to control of viral replication in a macaque AIDS model. An international collaborative project toward a clinical trial of an AIDS vaccine using SeV vectors is now proceeding. This vector with the potential to efficiently induce CTL as well as antibody responses would be useful for vaccines against a large variety of infectious diseases.

5.1 Introduction

Since the first attempt of a live-attenuated vaccine against smallpox in humans in 1798, many kinds of vaccines have been developed and contributed to the control of various infectious diseases. In addition to the traditional vaccines such as those live attenuated and inactivated, new technologies have been developed to obtain recombinant or purified protein vaccines, DNA vaccines, viral vector vaccines, and others. Sendai virus (SeV) vector is a novel vaccine delivery tool. The potential of vaccines using SeV vectors to control infectious diseases, including acquired immunodeficiency syndrome (AIDS) is discussed in this chapter.

S. Seki • T. Matano (✉)
AIDS Research Center, National Institute of Infectious Diseases,
1-23-1 Toyama, Shinjuku-ku, Tokyo 162-8640, Japan

The Institute of Medical Science, The University of Tokyo, 4-6-1 Shirokanedai,
Minato-ku, Tokyo 108-8639, Japan
e-mail: tmatano@nih.go.jp

Y. Nagai (ed.), *Sendai Virus Vector: Advantages and Applications*,
DOI 10.1007/978-4-431-54556-9_5,

5.2 Vaccine Technologies

Vaccines can be divided into two groups: one delivers proteins as antigens and the other delivers genes to express antigens. The former, including inactivated and protein vaccines, can induce exogenous antigen presentation and tends to be used for antibody elicitation. The latter, including live-attenuated, DNA, and viral vector vaccines, can induce endogenous as well as exogenous antigen presentation and is considered advantageous for elicitation of T-cell as well as antibody responses.

5.2.1 Live-Attenuated Vaccines

Live-attenuated vaccines are created by attenuating infectious microorganisms. They are capable of replicating but lose the potential to develop diseases. Live-attenuated vaccines have the potential to induce broad and long-lasting immune responses. Attenuation has been achieved by multiple rounds of microorganism proliferation in cell culture or genetic manipulation of virulent microorganisms. Naturally occurring avirulent variants are also used. Live-attenuated vaccines have a long history of successful use against a large variety of infectious diseases including smallpox (first used in 1798; the pathogen was eradicated in the late 1970s), tuberculosis (developed in 1921), yellow fever (developed in 1932), polio (developed in 1962), measles (developed in 1963), mumps (developed in 1967), rubella (developed in 1969), varicella (developed in 1974), typhoid (developed in 1983), Japanese encephalitis (developed in 1988), influenza (developed in 2003), herpes zoster (developed in 2005), and rotavirus infection (developed in 2006). However, we should be careful about potential risks of reversion to virulent form in vaccination with live-attenuated organisms.

5.2.2 Inactivated Vaccines

Inactivated vaccines use as antigens the whole pathogens, inactivated mostly by heat or chemical agents such as formalin and phenol. They have no ability to develop disease or proliferate in the host. Inactivated vaccines are considered advantageous for antibody induction but are not generally so effective for eliciting cellular immune responses, which are important in the control of intracellular infections. Multiple doses are usually required. Inactivated vaccines have also been used against a large variety of infectious diseases such as rabies (developed in 1885), typhoid (developed in 1886), cholera (developed in 1896), plague (developed in 1897), pertussis (developed in 1914), Japanese encephalitis (developed in 1935), influenza (developed in 1936), tick-borne encephalitis (developed in 1937), mumps (developed in 1948), polio (developed in 1955), and hepatitis A (developed in 1995). It is very important to strictly ensure complete inactivation of vaccine pathogens.

5.2.3 *Recombinant/Purified Protein Vaccines*

Advances in molecular biological technologies have enabled us to obtain recombinant microorganism proteins synthesized by heterologous virus, bacteria, or yeast expression systems and large amounts of proteins purified from pathogenic microorganisms. These recombinant or purified proteins are used for vaccine antigens mainly to induce humoral immune response. Protein vaccines are usually administered multiple times with adjuvants to enhance immune responses. These vaccines can be refined by target antigen selection and optimization, although it is often difficult to purify proteins with correct folding and optimized structure. A merit of this system is that there is no risk of microorganism proliferation. Recombinant or purified protein vaccines against the following infectious diseases are known: diphtheria (developed in 1923), tetanus (developed in 1926), pertussis (developed in 1981), hepatitis B (developed in 1986), cholera (developed in 1991), and human papillomavirus infection (developed in 2006).

Some pathogenic bacteria are covered with polysaccharide capsules that can be used for vaccine conjugates with carrier proteins to enhance phagocytosis and uptake of antigens by host antigen-presenting cells. These polysaccharide protein conjugate vaccines have been developed against infection with bacteria such as *Haemophilus influenzae* type b (developed in 1987), meningococcal (developed in 1999), pneumococcal (developed in 2000), and typhoid (developed in 2008).

5.2.4 *DNA Vaccines*

There have been many attempts to use DNA plasmids that express antigen proteins for vaccines. These DNA vaccines have the potential to elicit cellular as well as humoral immune responses. It is easy to prepare vaccine DNA plasmids, which is one of the merits of this system. This method is considered safe, but we need to be careful about the potential risk of DNA integration into the host chromosomes, although with very low frequencies. DNA vaccines have lower immunogenicity compared to viral vector vaccines expressing antigens but can be administered multiple times. The potential of DNA vaccines in priming has been recognized, and prime-boost vaccines are under development using DNA vaccines for priming in combination with other technologies such as viral vector vaccines for boosting. Some of them have proceeded to clinical trials.

5.2.5 *Viral Vector Vaccines*

Viruses can introduce viral genomes into the host cells by infection and express viral proteins efficiently. Viral vector vaccines expressing microorganism antigens

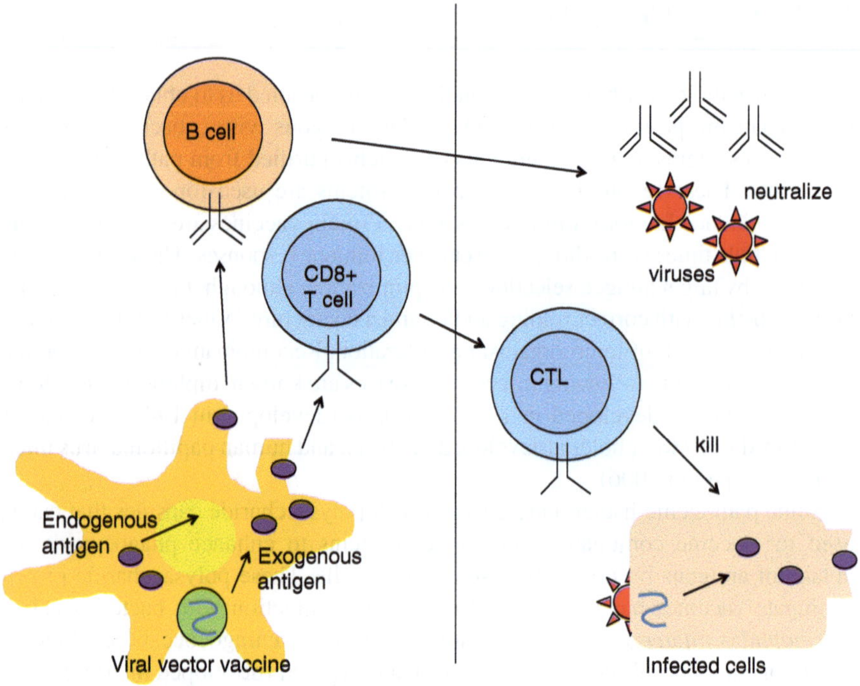

Fig. 5.1 Antigen presentation and induction of antibody and cytotoxic T-lymphocyte (CTL) responses by viral vector vaccination

are constructed by incorporation of engineered viral genomes inserted with antigen-coding genes into virions. Replication-competent vectors derived from avirulent or attenuated viruses, or replication-incompetent vectors, are used to ensure safety. Direct infection of host dendritic cells (DCs) with viral vectors would result in efficient antigen presentation via the endogenous pathway. Alternatively, viral vector-infected cells or antigens derived from those cells can be incorporated by DCs, leading to efficient antigen presentation by the exogenous pathway (Fig. 5.1). The potential of viral vector vaccines to efficiently induce cellular as well as humoral immune responses has been recognized. Especially, clinical trials of viral vector vaccines against malaria, tuberculosis, and AIDS have shown efficient induction of cellular immune responses (McShane et al. 2004; Bejon et al. 2006; Webster et al. 2006; Greenough et al. 2008). Viral vector vaccines are considered promising in both priming and boost. Most vectors are derived from viruses that can efficiently infect humans, possibly resulting in a high prevalence of people with anti-vector antibodies by natural infection. Such preexisting antibodies against vector viruses can be an obstacle to the immunogenicity of viral vector vaccines.

5.3 SeV Vector Vaccines

The SeV vector, with the potential to efficiently express antigens, is a promising vaccine delivery tool. This vector is considered superior in both safety and immunogenicity as vaccines.

5.3.1 Safety and Immunogenicity

SeV is a negative strand RNA virus, and the entire life cycle is completed in the cytoplasm using its own RNA polymerase. Thus, SeV vector infection has neither nuclear phase nor genotoxicity, which is superior to widely used adenovirus (AdV), adeno-associated virus, and retrovirus vectors with potential risk of genotoxicity from entry of viral genomes into the host nucleus (see Chaps. 2 and 3).

This virus causes a serious disease in rodents but is considered nonpathogenic in primates, including humans. A clinical trial phase I of SeV immunization as a live, xenotropic vaccine against human parainfluenza virus type 1 (hPIV-1) that can cross-react with anti-SeV antibodies showed no significant adverse outcomes, confirming the safety of intranasal SeV inoculation into humans (Slobod et al. 2004). Intranasal SeV inoculation of macaques shows transient SeV replication, which is localized around the nasal mucosa and peaks within a week. A macaque experiment revealed no efficient SeV transmission from SeV-infected to uninfected hosts (Kano et al. 2002).

Intranasal vaccination with SeV vectors can efficiently induce antigen-specific cytotoxic T-lymphocyte (CTL) responses in macaques (Kano et al. 2002; Matano et al. 2004). CTL induction is detected even in the tonsils, suggesting the potential of this vector to induce mucosal immune responses (Takeda et al. 2003). Not only replication-competent, transmissible but also nonpropagating, nontransmissible SeV vectors have been developed and are available for vaccines (Li et al. 2000; Takeda et al. 2008) (see Chap. 3).

5.3.2 Prevalence of People with SeV-Specific Neutralizing Antibodies

Natural infection with parental viruses induces antibody responses against vector viruses, which can be an obstacle to immune induction by viral vector vaccines. SeV vector vaccine is superior in this point, because its natural host is mice and there is no natural SeV infection in humans. Anti-hPIV-1 antibodies, however, may cross-react with SeV and affect the immunogenicity of this vector vaccination. Then, prevalence and titers of SeV-specific neutralizing antibodies (NAbs) in humans from Africa,

Europe, the United States, and Japan were analyzed previously (Hara et al. 2011). SeV NAb titers were determined by a sensitive virus neutralization assay using a SeV vector encoding a green fluorescent protein (GFP). A correlation was shown between anti-hPIV-1 antibody enzyme-linked immunosorbent assay (ELISA) and anti-SeV NAb titers, indicating that antibodies induced by natural infection with hPIV-1 can cross-react with SeV and neutralize it. Anti-SeV-neutralizing responses were detected in 92.5 % of subjects with a median titer of 60.6 (of 50 % inhibition). Titers were mostly less than 100 (71.7 %) and were greater than 1,000 in only 3.2 %. Thus, anti-SeV-neutralizing responses are detectable in the vast majority of adult populations but the titers are relatively low.

5.3.3 Influence of Preexisting Antivector Antibodies on Immunogenicity

We examined the influence of preexisting antivector antibodies on antigen-specific CTL induction by SeV vector vaccination in macaques (Moriya et al. 2008). Intranasal SeV vector vaccination twice with an interval of more than a year showed efficient antigen-specific CTL induction in macaques even after the second vaccination in the presence of anti-SeV NAb responses. Antigen-specific CTL induction was observed even by immunization with lower doses of SeV vectors in the presence of anti-SeV-neutralizing titers of 100 (of 90 % inhibition) (Moriya et al. 2011), which is higher than those observed in most human individuals, as already described. These results suggest the potential of intranasal SeV vector vaccination to efficiently induce antigen-specific CTL responses, even in humans.

5.3.4 SeV Inoculation Routes

Not only intranasal but also intramuscular vaccination with SeV vectors can efficiently induce antigen-specific immune responses. Comparison in macaques has shown no significant difference in immunogenicity between intranasal and intramuscular SeV vector vaccination. Interestingly, however, antigen-specific CTL responses were efficiently induced by intranasal SeV vector vaccination but not by intramuscular in pre-SeV-infected macaques with anti-SeV NAbs (Moriya et al. 2011). Thus, both intranasal and intramuscular SeV vaccination can equivalently induce antigen-specific CTL responses in the absence of anti-SeV antibodies, whereas intranasal SeV vaccination is more immunogenic than intramuscular in the presence of anti-SeV NAbs. Further studies have indicated enhancement of antigen-specific CTL responses by repeated SeV vector vaccination with short intervals of 3 weeks (Kurihara et al. 2012).

5.4 AIDS Vaccine Development

In the UNAIDS estimation, 34.0 million people are living with human immunodeficiency virus (HIV) in the world, and there were 2.7 million people newly infected with HIV and 1.8 million who died of AIDS in 2010 (UNAIDS 2011). A comprehensive strategy in combination with social prevention activities, vaccines, and therapeutics would be required for HIV control. Social prevention activities including education may be the most important but are not sufficient for the control of chronic and persistent HIV infection. Although potent anti-HIV therapy has dramatically reduced the morbidity and mortality of HIV infection, it is difficult to eradicate the virus from the host and cure HIV infection. A recent case study has suggested HIV eradication from an individual after stem cell transplantation with a mutant coreceptor (CCR5 delta 32 deletion) (Hütter et al. 2009). However, in most HIV-infected people, the virus is not eradicated and diversifies during the infection, with selection of multiple viral genome mutations resulting in escape from T-cell or antibody responses (McMichael 1998; Richman et al. 2003; Wei et al. 2003; Trkola et al. 2005). Thus, to avoid AIDS progression, HIV-infected people need to receive lifelong anti-retroviral therapy, which has many problems including tremendous cost, potential long-term toxicity, drug–drug interactions, difficulties in adherence, and risk of drug resistance. Development of a safe and effective preventive AIDS vaccine is thus desired for global control of HIV prevalence.

Despite great efforts to develop AIDS vaccines, we have not yet obtained an effective vaccine. The most essential feature of HIV infection is that the host immune responses fail to clear the virus and allow chronic infection in the natural course, indicating no spontaneous recovery (Fig. 5.2). Therefore, in contrast to traditional vaccines mimicking natural infection to induce immune responses leading to spontaneous control, AIDS vaccines need to induce immune memory leading to more effective immune responses than those induced by natural infection. Because some immune responses may not be effective but rather enhance HIV replication by activating target T lymphocytes, it is important to determine which immune memory induction can result in HIV control. Induction of effective T-cell and antibody responses is believed to be a promising strategy.

CTL responses are crucial for HIV control (Koup et al. 1994; Borrow et al., 1994; Matano et al. 1998), and HIV-specific CTL induction is an important strategy for development of an effective AIDS vaccine. In the early 2000s, several vaccine trials inducing T-cell responses showed protection from challenge with simian-human immunodeficiency virus SHIV89.6P in rhesus macaques (Barouch et al. 2000; Amara et al. 2001; Belyakov et al. 2001; Matano et al. 2001; Rose et al. 2001, Shiver et al. 2002). The SHIV89.6P, a chimeric simian immunodeficiency virus (SIV) whose env is replaced with HIV env, uses CXCR4 as the coreceptor and develops an acute AIDS with a rapid decrease in peripheral CD4+ T cells in macaques after challenge. This progression looks highly pathogenic, but later analyses indicated that this acute infection can be controlled more easily compared to

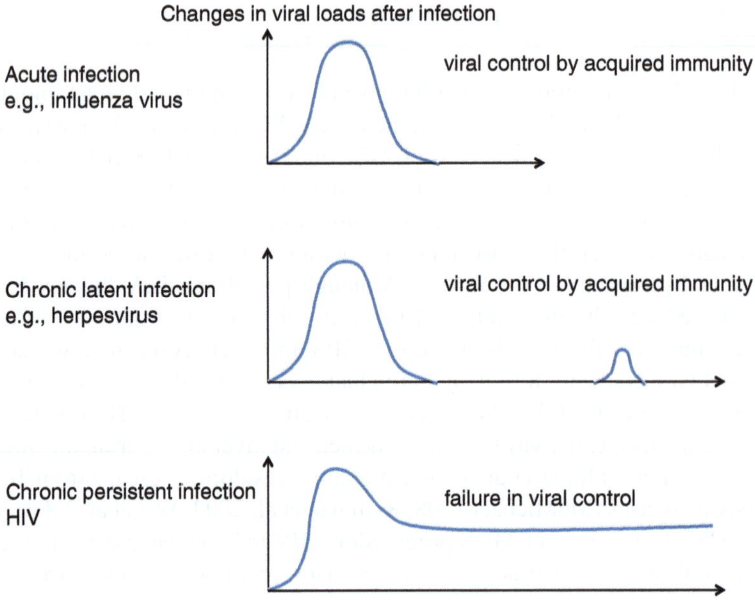

Fig. 5.2 Acute and latent viral infections with spontaneous control by acquired immune responses and persistent human immunodeficiency virus (HIV) infections without spontaneous control

chronic SIV infection. Indeed, T-cell-based vaccines that consistently showed SHIV89.6P control failed to show consistent SIV control, indicating that it is more difficult to control chronic infection with CCR5-tropic SIVs such as SIVmac239 (Horton et al. 2002). Thus, chronic SIV infection of macaques reflecting HIV infection in humans is currently considered as the best chronic AIDS model for evaluation of AIDS vaccine efficacy (Feinberg and Moore 2002).

There are now several vaccines that have shown protective efficacy in macaque chronic AIDS models for the first time, and all of them are T-cell-based vaccines that aim at antigen-specific T-cell induction by using viral vectors such as AdV and SeV (Matano et al. 2004; Letvin et al. 2006; Wilson et al. 2006; Liu et al. 2009). They may not result in sterile protection but do provide reduction of viral loads, possibly contributing to the control of HIV prevalence, and clinical trials of these vaccines are currently proceeding.

Broadly reactive HIV-specific NAb induction is believed to be critical for sterile HIV protection. However, inefficient NAb induction is a characteristic of primary HIV infection, and it is not easy to induce broadly reactive NAb responses. Epitopes targeted by several broadly reactive NAbs such as VRC01 and PG9 have been determined, and greater efforts have been made toward molecular design for NAb-inducing AIDS vaccines (Burton et al. 2010; Zhou et al. 2010).

Several vaccines using HIV Env as antigens have been attempted in clinical efficacy trials, although they were unable to induce anti-HIV NAb responses. Although the first clinical trial phase III of an Env gp120 vaccine (AIDSVAX B/E gp120) failed to show efficacy (Gilbert et al. 2005, Pitisuttithum et al. 2006), the RV144 trial of a prime-boost regimen combining the AIDSVAX B/E gp120 vaccine with an ALVAC canarypox vaccine has recently shown some efficacy, 30 % reduction in HIV prevalence by vaccination compared to the placebo, for the first time in Thailand (Rerks-Ngarm et al. 2009). Association of Env-specific IgG levels with HIV protection has been indicated (Haynes et al. 2012). Although efficacy was limited and transient, this study proved the possibility of vaccines to elicit protective immunity against HIV infection.

5.4.1 Antigen-Specific Antibody Responses

The HIV Env is a trimer of heterodimers consisting of gp120 (SU, surface protein) and gp41 (TM, transmembrane protein) and mediates virus entry into target cells by binding of gp120 to CD4 molecules and chemokine receptors (CKRs) on the cell surface. Binding to CD4 causes conformational changes in gp120, leading to the exposure and/or formation of a binding site for specific CKR, mainly CCR5 and CXCR4. NAbs play a central role in prevention of most viral infections, but Env-specific NAb induction is not efficient in the early phase of HIV infection. The unusual neutralization-resistant nature of HIV Env, such as variable loops, condensed N-linked glycosylation, and conformational flexibility of gp120, explain why the virus can easily evade the host's humoral immunity (Wyatt et al. 1998; Myszka et al. 2000; Kwong et al. 2002; Burton et al. 2004; Chen et al. 2009). A challenge experiment of macaques with a mutant SIV lacking some sites for N-linked glycosylation in Env suggested involvement of N-linked glycosylation in neutralization resistance (Reitter et al. 1998). HIV infection shows rapid selection of escape mutations in viral env, indicating some suppressive pressure on HIV replication by HIV-specific antibody responses (Richman et al. 2003; Wei et al. 2003; Trkola et al. 2005). Pre-challenge passive NAb immunization in macaque AIDS models has indicated the potential of NAbs to protect viral infection *in vivo* (Mascola et al. 1999; Mascola et al. 2000; Shibata et al. 1999; Parren et al. 2001; Veazey et al. 2003).

5.4.1.1 A gp120 Vaccine

In contrast to HIV strains adapted to laboratory culture, primary HIV isolates from infected individuals are resistant to neutralization (Moore et al. 1995). Thus, it is difficult to induce protective anti-HIV NAb responses (Berman et al. 1997), although several studies in the 1990s showed efficacy of vaccines inducing Env-specific antibodies against homologous HIV challenge in chimpanzees (Berman et al. 1990; Emini et al. 1992; Fultz 1992; Girard et al. 1995; Girard et al. 1997; Lubeck et al. 1997). However, a vaccine using gp120 as antigens was tested in the first phase III efficacy trial for an AIDS vaccine (Flynn et al. 2005). Using two recombinant gp120 proteins derived from two different HIV clade B strains as immunogens, the vaccine elicited no effective NAb responses and failed to show efficacy.

5.4.1.2 Broadly Reactive NAb Responses for HIV Protection

Analyses of neutralizing responses in sera from HIV-infected people have detected several monoclonal antibodies that can neutralize primary HIV isolates *in vitro* (Moore et al. 2001). Recently, broadly reactive, potently neutralizing sera have been found in some HIV-infected individuals (Li et al. 2007; Stamatatos et al. 2009; Walker et al. 2010). The human immune system takes several years to generate such broadly reactive NAbs. It is now possible to isolate and characterize these antibodies with new technologies (Scheid et al. 2011; Walker et al. 2011; Wu et al. 2011), contributing to our understanding of the process for broadly reactive NAb evolution. Now, many basic attempts are proceeding to make a molecular design for inducing these broadly reactive NAb responses.

5.4.1.3 Contribution of Post-Exposure NAb Responses to HIV Control

Induction of broadly reactive NAb memory would contribute to HIV protection, but technical difficulties in achieving requisite neutralizing titers for sterile HIV protection have been suggested. Several post-challenge passive NAb immunization studies in macaques suggested that NAb induction could have only a limited suppressive effect on primary HIV replication once infection is established (Haigwood et al. 1996; Nishimura et al. 2003). However, a recent study of passive NAb immunization 1 week after SIV challenge showed significant reduction in set-point viral loads (Yamamoto et al. 2007). Further analyses indicated NAb-triggered induction of polyfunctional CD4+ T-cell responses, possibly by enhancement of NAb-mediated DC uptake of virions (Yamamoto et al. 2009; Yamamoto and Matano 2010). Thus, even substerile titers of NAbs can contribute to primary HIV control.

5.4.2 *Antigen-Specific T-Cell Responses*

CTLs recognize HIV-infected cells by specific detection of HIV antigen-derived epitopes (peptides)–MHC-I complexes presented on the cell surface (Townsend et al. 1986) and kill the infected cells. T-cell-based AIDS vaccines are aiming at efficient induction of effective T-cell responses, leading to control of HIV replication post exposure. Optimization of antigens and delivery systems is important. Viral vectors are promising vaccine delivery tools, and several vaccines using viral vectors have shown efficacy in macaque chronic AIDS models of SIV infection. The potency of these T-cell-based vaccines would be enhanced by antigen optimization and combination with multiple delivery systems.

5.4.2.1 Viral Vector AIDS Vaccines

Since the mid-1990s, the potency of a variety of viral vectors as AIDS vaccines has been tested, such as canarypox virus (Egan et al. 1995), fowlpox virus (Kent et al. 1998), modified vaccinia Ankara virus (MVA) (Hanke et al. 1999), Semliki Forest virus (Mossman et al. 1996), Venezuelan equine encephalitis virus (Caley et al. 1997), cytomegalovirus (CMV) (Hansen et al. 2011), adeno-associated virus (Xin et al. 2001), AdV (Shiver et al. 2002), and SeV (Kano et al. 2000; Matano et al. 2001) vectors. Prime-boost regimens consisting of multiple viral vectors or DNA vaccines are considered the best to induce efficient antigen-specific T-cell responses. Although a DNA vaccine alone can induce weak T-cell responses, boosting by a viral vector vaccine elicits stronger responses (Hanke et al. 1999; Allen et al. 2000). Especially, prime-boost regimens using AdV, MVA, or SeV vectors have shown significant reduction in viral loads in macaque AIDS models of SIV infection (Matano et al. 2004; Goulder and Watkins 2004; Casimiro et al. 2005; Letvin et al. 2006; Wilson et al. 2006; Kawada et al. 2007). Most T-cell-based vaccines aim at T-cell memory induction leading to effective T-cell responses to control HIV replication post exposure, whereas CMV vectors have the potential to elicit CTL effectors, possibly contributing to protection of HIV infection (Hansen et al. 2011).

The clinical efficacy trial of a vaccine using replication-incompetent AdV type 5 (Ad5) vectors expressing HIV Gag, Pol, and Nef proteins failed to show efficacy and revealed the issues on prevalence of anti-Ad5 antibodies (Buchbinder et al. 2008; Duerr et al. 2012). Now, engineered AdV vectors resistant to anti-Ad5 antibodies and other types of AdV vectors with low prevalence of antivector antibodies have been developed (Liu et al. 2009). Antivector antibodies can also be an obstacle to the potency of repeated viral vector vaccination. Thus, combination of heterologous viral vector vaccines is now an important strategy for AIDS vaccine development.

5.4.2.2 SeV Vector-Based AIDS Vaccines

Our studies have shown safety and efficacy of prophylactic DNA-prime/SeV vector-boost AIDS vaccines in macaque AIDS models. Prime-boost regimens using ΔFSeV as well as full length SeV vectors expressing Gag antigen (see Chap. 3) elicited efficient Gag-specific CTL responses and showed consistent protection of SHIV89.6P infection (Matano et al. 2001; Takeda et al. 2003). In a chronic AIDS model, five of eight animals that received this prime-boost vaccine controlled SIVmac239 replication, indicating the protective efficacy of T-cell-based vaccines against pathogenic SIV challenge for the first time (Matano et al. 2004; Kawada et al. 2007). An international collaborative clinical trial of SeV vector-based AIDS vaccine was started in 2013 (http://www.iavi.org/Information-Center/Press-Releases/Pages/IAVI-AND-PARTNERS-INITIATE-PHASE-I-TRIAL-OF-A-NOVEL-AIDS-VACCINE-REGIMEN.aspx).

These vaccine-based SIV controllers showed rapid selection of viral gag CTL escape mutations with viral fitness costs, suggesting strong suppressive pressure on

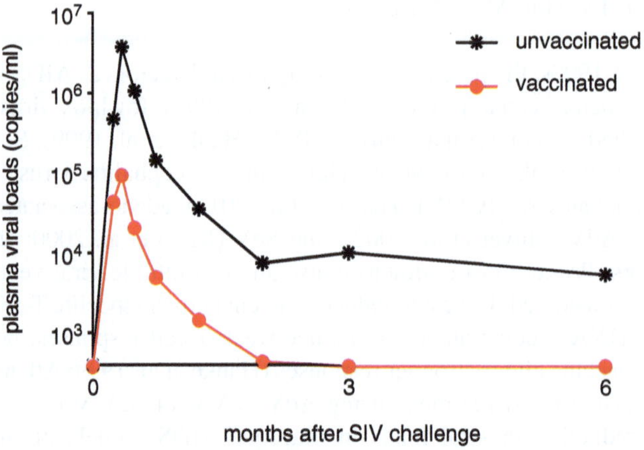

Fig. 5.3 Changes in geometric means of plasma viral loads (SIV gag RNA copies/ml plasma) in unvaccinated (*black*) and DNA-prime/SeV-Gag boost-vaccinated (*red*) macaques possessing the MHC-I haplotype 90-120-Ia after SIVmac239 challenge. All the vaccinated animals controlled simian immunodeficiency virus (SIV) replication. (Modified from fig. 3B in Ishii et al. 2012)

SIV replication by Gag-specific CTL responses. In a group of Burmese rhesus macaques [*Rhesus (Macaca) mulatta*] sharing MHC-I haplotype 90-120-Ia (Naruse et al. 2010; Nomura et al. 2012), all the vaccinees controlled a SIVmac239 challenge and showed undetectable set-point viremia (Fig. 5.3). In contrast, all the vaccinated animals possessing 90-120-Ia failed to control a challenge with a mutant SIV carrying gag CTL escape mutations, indicating that these Gag-specific CTL responses are responsible for the vaccine-based SIV control (Kawada et al. 2006; Kawada et al. 2008).

A study on a vaccine using SeV vectors expressing a single CTL epitope has indicated that even a CTL memory response to a single epitope that can recognize the incoming diversified HIVs could facilitate HIV control (Tsukamoto et al. 2009). Furthermore, this study has demonstrated that virus-specific CTL memory induced by vaccination without virus-specific CD4$^+$ T-cell help could facilitate SIV control after virus exposure, indicating the benefit of prophylactic vaccination eliciting virus-specific CTL memory with non-virus-specific CD4$^+$ T-cell responses for HIV control.

5.4.2.3 Antigen Optimization

In a clinical trial, Ad5 vector vaccination induced T-cell responses directed against only limited numbers of epitopes (Rolland et al. 2011). Thus, it is considered important to elicit broad CTL responses, although broadening of CTL responses may not always result in positive outcomes. Mosaic vaccines, the polyvalent type that aim at inducing immune responses which can recognize genetically diverse viruses, have been developed for induction of broad T-cell responses (Fischer et al. 2007;

Barouch et al. 2010; Santra et al. 2010). Immunogens derived from natural HIV strains are optimized in silico to provide maximal coverage of potential T-cell epitopes. On the other hand, conserved epitopes including dozens of highly conserved viral genome segments have been developed (Létourneau et al. 2007; Rolland et al. 2007). Conserved epitopes are identified in HIV Gag, Pol, Env, and Vif proteins.

In particular, the contribution of CTL responses targeting Gag epitopes to HIV control has been indicated (Zuñiga et al. 2006; Kiepiela et al. 2007). A recent study using SeV vectors presents evidence indicating significant influence of prophylactic vaccination on postexposure CTL immunodominance and cooperation of vaccine antigen-specific and non-vaccine antigen-specific CTL responses, which affects virus control (Ishii et al. 2012).

5.5 SeV Vector Vaccines Against Other Infectious Diseases

Infection with the protozoan parasite *Trypanosoma cruzi* (*T. cruzi*) causes Chagas' disease, which was first recognized by the Brazilian doctor Carlos Chagas in 1909. It is estimated that 15–16 million inhabitants are currently infected with *T. cruzi* in Latin America and 75–90 million people are exposed to infection (Coura 2007). *T. cruzi* in bug feces can infect humans by entering the body through the mucosa. It can also be transmitted by blood transfusion or organ transplant. *T. cruzi* infection via the bloodstream and lymphatics is established in muscles and cardiac tissues. There are two phases of Chagas' disease: the acute and the chronic phases. The acute phase lasts for the first several weeks or months of infection. It usually goes unnoticed because infected people have no or mild symptoms. However, approximately 30 % of infected individuals will develop complications in their cardiac, digestive, or peripheral nervous systems in the chronic phase after decades of infection. Because the therapeutic drugs are not effective in the chronic phase, treatment in the early phase is important for avoiding lifelong and life-threatening infection. Chagas' disease results in loss of labor productivity consequent to premature disability and death. It is thus necessary to develop an effective vaccine for controlling this disease.

Similar to other parasites that invade the bloodstream such as the preerythrocytic stages of *Plasmodium* spp. and *Toxoplasma gondii*, CTL and interferon-γ responses are considered effective against *T. cruzi* infection. Several studies have been conducted to develop a vaccine using live-attenuated (Menezes 1968) or inactivated *T. cruzi* in animal models (Basombrio 1990). However, these vaccines could only delay or decrease disease onset. To effectively activate CTL responses, DNA vaccines have also been attempted with some of the most promising antigens (Costa et al. 1998; Wizel et al. 1998; Fujimura et al. 2001; Garg and Tarleton 2002; Katae et al. 2002; Boscardin et al. 2003; Fralish and Tarleton 2003; Dumonteil et al. 2004; Vasconcelos et al. 2004; Cazorla et al. 2008; Chou et al. 2008). Viral vectors expressing such antigens are expected to induce stronger CTL responses than these DNA vaccines.

A ΔFSeV encoding the *T. cruzi* amastigote (proliferating form of the parasite) surface protein (ASP)-2 has recently been developed (Duan et al. 2009). Intranasal immunization of C57BL/6 mice with this vector displayed significantly lower parasitemia and higher survival rates after challenge. CD8 depletion confirmed the critical role of CTL responses in this control of *T. cruzi* infection. In addition, a single intranasal vaccination was sufficient, whereas DNA vaccines and other viral vector vaccines in previous studies (Garg and Tarleton 2002; Boscardin et al. 2003; Vasconcelos et al. 2004; Machado et al. 2006; Chou et al. 2008) required multiple vaccinations to induce protective immunity against *T. cruzi*, indicating higher immunogenicity of the SeV vector vaccine.

SeV vectors may also be a promising vaccine tool against respiratory syncytial virus, which causes serious lower respiratory tract infections in young children. For details, see Chap. 4.

5.6 Conclusion

Vaccine development is still challenging against a variety of infectious diseases. Traditional vaccines mostly aim at effective antibody responses, whereas T-cell induction would be effective especially against persistent infectious diseases. SeV vectors are promising vaccine delivery tools for efficient induction of T-cell as well as antibody responses. The combination of SeV vector vaccines with other vaccines could lead to development of an optimal effective vaccine.

References

Allen TM, Vogel TU, Fuller DH, Mothé BR, Steffen S, Boyson JE, Shipley T, Fuller J, Hanke T, Sette A, Altman JD, Moss B, McMichael AJ, Watkins DI (2000) Induction of AIDS virus-specific CTL activity in fresh, unstimulated peripheral blood lymphocytes from rhesus macaques vaccinated with a DNA prime/modified vaccinia virus Ankara boost regimen. J Immunol 164(9):4968–4978, http://www.jimmunol.org/content/164/9/4968.full.pdf+html

Amara RR, Villinger F, Altman JD, Lydy SL, O'Neil SP, Staprans SI, Montefiori DC, Xu Y, Herndon JG, Wyatt LS, Candido MA, Kozyr NL, Earl PL, Smith JM, Ma HL, Grimm BD, Hulsey ML, Miller J, McClure HM, McNicholl JM, Moss B, Robinson HL (2001) Control of a mucosal challenge and prevention of AIDS by a multiprotein DNA/MVA vaccine. Science 292:69–74. doi:10.1126/science.292.5514.69

Barouch DH, Santra S, Schmitz JE, Kuroda MJ, Fu TM, Wagner W, Bilska M, Craiu A, Zheng XX, Krivulka GR, Beaudry K, Lifton MA, Nickerson CE, Trigona WL, Punt K, Freed DC, Guan L, Dubey S, Casimiro D, Simon A, Davies ME, Chastain M, Strom TB, Gelman RS, Montefiori DC, Lewis MG, Emini EA, Shiver JW, Letvin NL (2000) Control of viremia and prevention of clinical AIDS in rhesus monkeys by cytokine-augmented DNA vaccination. Science 290:486–492. doi:10.1126/science.290.5491.486

Barouch DH, O'Brien KL, Simmons NL, King SL, Abbink P, Maxfield LF, Sun YH, La Porte A, Riggs AM, Lynch DM, Clark SL, Backus K, Perry JR, Seaman MS, Carville A, Mansfield KG, Szinger JJ, Fischer W, Muldoon M, Korber B (2010) Mosaic HIV-1 vaccines expand the breadth and depth of cellular immune responses in rhesus monkeys. Nat Med 16(3):319–323. doi:10.1038/nm.2089

Basombrio MA (1990) *Trypanosoma cruzi*: partial prevention of the natural infection of guinea pigs with a killed parasite vaccine. Exp Parasitol 71(1):1–8. doi:10.1016/0014-4894(90)90002-T

Bejon P, Mwacharo J, Kai OK, Todryk S, Keating S, Lang T, Gilbert SC, Peshu N, Marsh K, Hill AV (2006) Immunogenicity of the candidate malaria vaccines FP9 and modified vaccinia virus Ankara encoding the pre-erythrocytic antigen ME-TRAP in 1- to 6-year-old children in a malaria endemic area. Vaccine 24(22):4709–4715. doi:10.1016/j.vaccine.2006.03.029

Belyakov IM, Hel Z, Kelsall B, Kuznetsov VA, Ahlers JD, Nacsa J, Watkins DI, Allen TM, Sette A, Altman J, Woodward R, Markham PD, Clements JD, Franchini G, Strober W, Berzofsky JA (2001) Mucosal AIDS vaccine reduces disease and viral load in gut reservoir and blood after mucosal infection of macaques. Nat Med 7:1320–1326. doi:10.1038/nm1201-1320

Berman PW, Gregory TJ, Riddle L, Nakamura GR, Champe MA, Porter JP, Wurm FM, Hershberg RD, Cobb EK, Eichberg JW (1990) Protection of chimpanzees from infection by HIV-1 after vaccination with recombinant glycoprotein gp120 but not gp160. Nature (Lond) 345:622–625. doi:10.1038/345622a0

Berman PW, Gray AM, Wrin T, Vennari JC, Eastman DJ, Nakamura GR, Francis DP, Gorse G, Schwartz DH (1997) Genetic and immunologic characterization of viruses infecting MN-rgp120-vaccinated volunteers. J Infect Dis 176:384–397. doi:10.1086/514055

Borrow P, Lewicki H, Hahn BH, Shaw GM, Oldstone MB (1994) Virus-specific CD8+ cytotoxic T-lymphocyte activity associated with control of viremia in primary human immunodeficiency virus type 1 infection. J Virol 68:6103–6110, http://jvi.asm.org/content/68/9/6103.long

Boscardin SB, Kinoshita SS, Fujimura AE, Rodrigues MM (2003) Immunization with cDNA expressed by amastigotes of *Trypanosoma cruzi* elicits protective immune response against experimental infection. Infect Immun 71(5):2744–2757. doi:10.1128/IAI.71.5.2744-2757.2003

Buchbinder SP, Mehrotra DV, Duerr A, Fitzgerald DW, Mogg R, Li D, Gilbert PB, Lama JR, Marmor M, Del Rio C, McElrath MJ, Casimiro DR, Gottesdiener KM, Chodakewitz JA, Corey L, Robertson MN, Step Study Protocol Team (2008) Efficacy assessment of a cell-mediated immunity HIV-1 vaccine (the step study): a double-blind, randomised, placebo-controlled, test-of-concept trial. Lancet 372(9653):1881–1893. doi:10.1016/S0140-6736(08)61591-3

Burton DR, Desrosiers RC, Doms RW, Koff WC, Kwong PD, Moore JP, Nabel GJ, Sodroski J, Wilson IA, Wyatt RT (2004) HIV vaccine design and the neutralizing antibody problem. Nat Immunol 5:233–236. doi:10.1038/ni0304-233

Burton DR et al (2010) Variable loop glycan dependency of the broad and potent HIV-1-neutralizing antibodies PG9 and PG16. J Virol 84:10510–10521

Caley IJ, Betts MR, Irlbeck DM, Davis NL, Swanstrom R, Frelinger JA, Johnston RE (1997) Humoral, mucosal, and cellular immunity in response to a human immunodeficiency virus type 1 immunogen expressed by a Venezuelan equine encephalitis virus vaccine vector. J Virol 71:3031–3038, http://jvi.asm.org/content/71/4/3031.long

Casimiro DR, Wang F, Schleif WA, Liang X, Zhang ZQ, Tobery TW, Davies ME, McDermott AB, O'Connor DH, Fridman A, Bagchi A, Tussey LG, Bett AJ, Finnefrock AC, Fu TM, Tang A, Wilson KA, Chen M, Perry HC, Heidecker GJ, Freed DC, Carella A, Punt KS, Sykes KJ, Huang L, Ausensi VI, Bachinsky M, Sadasivan-Nair U, Watkins DI, Emini EA, Shiver JW (2005) Attenuation of simian immunodeficiency virus SIVmac239 infection by prophylactic immunization with DNA and recombinant adenoviral vaccine vectors expressing gag. J Virol 79(24):15547–15555. doi:10.1128/JVI.79.24.15547-15555.2005

Cazorla SI, Frank FM, Becker PD, Corral RS, Guzmán CA, Malchiodi EL (2008) Prime-boost immunization with cruzipain co-administered with MALP-2 triggers a protective immune response able to decrease parasite burden and tissue injury in an experimental *Trypanosoma cruzi* infection model. Vaccine 26(16):1999–2009. doi:10.1016/j.vaccine.2008.02.011

Chen L, Kwon YD, Zhou T, Wu X, O'Dell S, Cavacini L, Hessell AJ, Pancera M, Tang M, Xu L, Yang ZY, Zhang MY, Arthos J, Burton DR, Dimitrov DS, Nabel GJ, Posner MR, Sodroski J, Wyatt R, Mascola JR, Kwong PD (2009) Structural basis of immune evasion at the site of CD4 attachment on HIV-1 gp120. Science 326:1123–1127. doi:10.1126/science.1175868

Chou B, Hisaeda H, Shen J, Duan X, Imai T, Tu L, Murata S, Tanaka K, Himeno K (2008) Critical contribution of immunoproteasomes in the induction of protective immunity against

Trypanosoma cruzi in mice vaccinated with a plasmid encoding a CTL epitope fused to green fluorescence protein. Microbes Infect 10(3):241–250. doi:10.1016/j.micinf.2007.11.010

Costa F, Franchin G, Pereira-Chioccola VL, Ribeirão M, Schenkman S, Rodrigues MM (1998) Immunization with a plasmid DNA containing the gene of trans-sialidase reduces *Trypanosoma cruzi* infection in mice. Vaccine 16(8):768–774. doi:10.1016/S0264-410X(97)00277-6

Coura JR (2007) Chagas disease: what is known and what is needed–a background article. Mem Inst Oswaldo Cruz 102(suppl 1):113–122. doi:10.1590/S0074-02762007000900018

Duan X, Yonemitsu Y, Chou B, Yoshida K, Tanaka S, Hasegawa M, Tetsutani K, Ishida H, Himeno K, Hisaeda H (2009) Efficient protective immunity against *Trypanosoma cruzi* infection after nasal vaccination with recombinant Sendai virus vector expressing amastigote surface protein-2. Vaccine 27(44):6154–6159. doi:10.1016/j.vaccine.2009.08.026

Duerr A, Huang Y, Buchbinder S, Coombs RW, Sanchez J, Del Rio C, Casapia M, Santiago S, Gilbert P, Corey L, Robertson MN, Step/HVTN 504 Study Team (2012) Extended follow-up confirms early vaccine-enhanced risk of HIV acquisition and demonstrates waning effect over time among participants in a randomized trial of recombinant adenovirus HIV vaccine (step study). J Infect Dis 206(2):258–266. doi:10.1093/infdis/jis342

Dumonteil E, Escobedo-Ortegon J, Reyes-Rodriguez N, Arjona-Torres A, Ramirez-Sierra MJ (2004) Immunotherapy of *Trypanosoma cruzi* infection with DNA vaccines in mice. Infect Immun 72(1):46–53. doi:10.1128/IAI.72.1.46-53.2004

Egan MA, Pavlat WA, Tartaglia J, Paoletti E, Weinhold KJ, Clements ML, Siliciano RF (1995) Induction of human immunodeficiency virus type 1 (HIV-1)-specific cytolytic T lymphocyte responses in seronegative adults by a nonreplicating, host-range-restricted canarypox vector (ALVAC) carrying the HIV-1MN.env gene. J Infect Dis 171:1623–1627. doi:10.1093/infdis/171.6.1623

Emini EA, Schleif WA, Nunberg JH, Conley AJ, Eda Y, Tokiyoshi S, Putney SD, Matsushita S, Cobb KE, Jett CM, Eichberg JW, Murthy KK (1992) Prevention of HIV-1 infection in chimpanzees by gp120 V3 domain-specific monoclonal antibody. Nature (Lond) 355:728–730. doi:10.1038/355728a0

Feinberg MB, Moore JP (2002) AIDS vaccine models: challenging challenge viruses. Nat Med 8(3):207–210. doi:10.1038/nm0302-207

Fischer W, Perkins S, Theiler J, Bhattacharya T, Yusim K, Funkhouser R, Kuiken C, Haynes B, Letvin NL, Walker BD, Hahn BH, Korber BT (2007) Polyvalent vaccines for optimal coverage of potential T-cell epitopes in global HIV-1 variants. Nat Med 13(1):100–106. doi:10.1038/nm1461

Flynn NM, Forthal DN, Harro CD, Judson FN, Mayer KH, Para MF, rgp120 HIV Vaccine Study Group (2005) Placebo-controlled phase 3 trial of a recombinant glycoprotein 120 vaccine to prevent HIV-1 infection. J Infect Dis 191(5):654–665. doi:10.1086/428404

Fralish BH, Tarleton RL (2003) Genetic immunization with LYT1 or a pool of trans-sialidase genes protects mice from lethal *Trypanosoma cruzi* infection. Vaccine 21(21–22):3070–3080. doi:10.1016/S0264-410X(03)00121-X

Fujimura AE, Kinoshita SS, Pereira-Chioccola VL, Rodrigues MM (2001) DNA sequences encoding CD4+ and CD8+ T-cell epitopes are important for efficient protective immunity induced by DNA vaccination with a *Trypanosoma cruzi* gene. Infect Immun 69(9):5477–5486. doi:10.1128/IAI.69.9.5477-5486.2001

Fultz PN (1992) Immunization and challenge of chimpanzees with HIV-1. AIDS Res Hum Retroviruses 8:1517–1519. doi:10.1089/aid.1992.8.1517

Garg N, Tarleton RL (2002) Genetic immunization elicits antigen-specific protective immune responses and decreases disease severity in *Trypanosoma cruzi* infection. Infect Immun 70(10):5547–5555. doi:10.1128/IAI.70.10.5547-5555.2002

Gilbert PB, Peterson ML, Follmann D, Hudgens MG, Francis DP, Gurwith M, Heyward WL, Jobes DV, Popovic V, Self SG, Sinangil F, Burke D, Berman PW (2005) Correlation between immunologic responses to a recombinant glycoprotein 120 vaccine and incidence of HIV-1 infection in a phase 3 HIV-1 preventive vaccine trial. J Infect Dis 191(5):666–677. doi:10.1086/428405

Girard M, Meignier B, Barré-Sinoussi F, Kieny MP, Matthews T, Muchmore E, Nara PL, Wei Q, Rimsky L, Weinhold K (1995) Vaccine-induced protection of chimpanzees against infection by a heterologous human immunodeficiency virus type 1. J Virol 69:6239–6248, http://jvi.asm.org/content/69/10/6239.long

Girard M, van der Ryst E, Barré-Sinoussi F, Nara P, Tartaglia J, Paoletti E, Blondeau C, Jennings M, Verrier F, Meignier B, Fultz PN (1997) Challenge of chimpanzees immunized with a recombinant canarypox–HIV-1 virus. Virology 232:98–104. doi:10.1006/viro.1997.8560

Goulder PJ, Watkins DI (2004) HIV and SIV CTL escape: implications for vaccine design. Nat Rev Immunol 4(8):630–640. doi:10.1038/nri1417

Greenough TC, Cunningham CK, Muresan P, McManus M, Persaud D, Fenton T, Barker P, Gaur A, Panicali D, Sullivan JL, Luzuriaga K, Pediatric AIDS Clinical Trials Group P1059 Team (2008) Safety and immunogenicity of recombinant poxvirus HIV-1 vaccines in young adults on highly active antiretroviral therapy. Vaccine 26(52):6883–6893. doi:10.1016/j.vaccine.2008.09.084

Haigwood NL, Watson A, Sutton WF, McClure J, Lewis A, Ranchalis J, Travis B, Voss G, Letvin NL, Hu SL, Hirsch VM, Johnson PR (1996) Passive immune globulin therapy in the SIV/macaque model: early intervention can alter disease profile. Immunol Lett 51:107–114. doi:10.1016/0165-2478(96)02563-1

Hanke T, Samuel RV, Blanchard TJ, Neumann VC, Allen TM, Boyson JE, Sharpe SA, Cook N, Smith GL, Watkins DI, Cranage MP, McMichael AJ (1999) Effective induction of simian immunodeficiency virus-specific cytotoxic T lymphocytes in macaques by using a multiepitope gene and DNA prime-modified vaccinia virus Ankara boost vaccination regimen. J Virol 73(9):7524–7532, http://jvi.asm.org/content/73/9/7524.full.pdf+html

Hansen SG, Ford JC, Lewis MS, Ventura AB, Hughes CM, Coyne-Johnson L, Whizin N, Oswald K, Shoemaker R, Swanson T, Legasse AW, Chiuchiolo MJ, Parks CL, Axthelm MK, Nelson JA, Jarvis MA, Piatak M Jr, Lifson JD, Picker LJ (2011) Profound early control of highly pathogenic SIV by an effector memory T-cell vaccine. Nature (Lond) 473(7348):523–527. doi:10.1038/nature10003

Hara H, Hironaka T, Inoue M, Iida A, Shu T, Hasegawa M, Nagai Y, Falsey AR, Kamali A, Anzala O, Sanders EJ, Karita E, Mwananyanda L, Vasan S, Lombardo A, Parks CL, Sayeed E, Krebs M, Cormier E, Ackland J, Price MA, Excler JL (2011) Prevalence of specific neutralizing antibodies against Sendai virus in populations from different geographic areas: implications for AIDS vaccine development using Sendai virus vectors. Hum Vaccin 7(6):639–645. doi:10.4161/hv.7.6.15408

Haynes BF, Gilbert PB, McElrath MJ, Zolla-Pazner S, Tomaras GD, Alam SM, Evans DT, Montefiori DC, Karnasuta C, Sutthent R, Liao HX, DeVico AL, Lewis GK, Williams C, Pinter A, Fong Y, Janes H, DeCamp A, Huang Y, Rao M, Billings E, Karasavvas N, Robb ML, Ngauy V, de Souza MS, Paris R, Ferrari G, Bailer RT, Soderberg KA, Andrews C, Berman PW, Frahm N, De Rosa SC, Alpert MD, Yates NL, Shen X, Koup RA, Pitisuttithum P, Kaewkungwal J, Nitayaphan S, Rerks-Ngarm S, Michael NL, Kim JH (2012) Immune-correlates analysis of an HIV-1 vaccine efficacy trial. N Engl J Med 366(14):1275–1286, http://www.nejm.org/doi/full/10.1056/NEJMoa1113425

Horton H, Vogel TU, Carter DK, Vielhuber K, Fuller DH, Shipley T, Fuller JT, Kunstman KJ, Sutter G, Montefiori DC, Erfle V, Desrosiers RC, Wilson N, Picker LJ, Wolinsky SM, Wang C, Allison DB, Watkins DI (2002) Immunization of rhesus macaques with a DNA prime/modified vaccinia virus Ankara boost regimen induces broad simian immunodeficiency virus (SIV)-specific T-cell responses and reduces initial viral replication but does not prevent disease progression following challenge with pathogenic SIVmac239. J Virol 76:7187–7202. doi:10.1128/JVI.76.14.7187-7202.2002

Hütter G, Nowak D, Mossner M, Ganepola S, Müssig A, Allers K, Schneider T, Hofmann J, Kücherer C, Blau O, Blau IW, Hofmann WK, Thiel E (2009) Long-term control of HIV by CCR5 Delta32/Delta32 stem-cell transplantation. N Engl J Med 360(7):692–698, http://www.nejm.org/doi/full/10.1056/NEJMoa0802905

Ishii H, Kawada M, Tsukamoto T, Yamamoto H, Matsuoka S, Shiino T, Takeda A, Inoue M, Iida A, Hara H, Shu T, Hasegawa M, Naruse TK, Kimura A, Takiguchi M, Matano T (2012) Impact of vaccination on cytotoxic T lymphocyte immunodominance and cooperation against simian immunodeficiency virus replication in rhesus macaques. J Virol 86:738–745. doi:10.1128/JVI.06226-11

Kano M, Matano T, Nakamura H, Takeda A, Kato A, Ariyoshi K, Mori K, Sata T, Nagai Y (2000) Elicitation of protective immunity against simian immunodeficiency virus infection by a recombinant Sendai virus expressing the Gag protein. AIDS 14(9):1281–1282. doi:10.1097/00002030-200006160-00030

Kano M, Matano T, Kato A, Nakamura H, Takeda A, Suzaki Y, Ami Y, Terao K, Nagai Y (2002) Primary replication of a recombinant Sendai viral vector in macaques. J Gen Virol 83:1377–1386

Katae M, Miyahira Y, Takeda K, Matsuda H, Yagita H, Okumura K, Takeuchi T, Kamiyama T, Ohwada A, Fukuchi Y, Aoki T (2002) Coadministration of an interleukin-12 gene and a *Trypanosoma cruzi* gene improves vaccine efficacy. Infect Immun 70(9):4833–4840. doi:10.1128/IAI.70.9.4833-4840.2002

Kawada M, Igarashi H, Takeda A, Tsukamoto T, Yamamoto H, Dohki S, Takiguchi M, Matano T (2006) Involvement of multiple epitope-specific cytotoxic T-lymphocyte responses in vaccine-based control of simian immunodeficiency virus replication in rhesus macaques. J Virol 80(4):1949–1958. doi:10.1128/JVI.80.4.1949-1958.2006

Kawada M, Tsukamoto T, Yamamoto H, Takeda A, Igarashi H, Watkins DI, Matano T (2007) Long-term control of simian immunodeficiency virus replication with central memory CD4+ T-cell preservation after nonsterile protection by a cytotoxic T-lymphocyte-based vaccine. J Virol 81(10):5202–5211. doi:10.1128/JVI.02881-06

Kawada M, Tsukamoto T, Yamamoto H, Iwamoto N, Kurihara K, Takeda A, Moriya C, Takeuchi H, Akari H, Matano T (2008) Gag-specific cytotoxic T lymphocyte-based control of primary simian immunodeficiency virus replication in a vaccine trial. J Virol 82:10199–10206. doi:10.1128/JVI.01103-08

Kent SJ, Zhao A, Best SJ, Chandler JD, Boyle DB, Ramshaw IA (1998) Enhanced T-cell immunogenicity and protective efficacy of a human immunodeficiency virus type 1 vaccine regimen consisting of consecutive priming with DNA and boosting with recombinant fowlpox virus. J Virol 72:10180–10188, http://jvi.asm.org/content/72/12/10180.full.pdf+html

Kiepiela P, Ngumbela K, Thobakgale C, Ramduth D, Honeyborne I, Moodley E, Reddy S, de Pierres C, Mncube Z, Mkhwanazi N, Bishop K, van der Stok M, Nair K, Khan N, Crawford H, Payne R, Leslie A, Prado J, Prendergast A, Frater J, McCarthy N, Brander C, Learn GH, Nickle D, Rousseau C, Coovadia H, Mullins JI, Heckerman D, Walker BD, Goulder P (2007) CD8+ T-cell responses to different HIV proteins have discordant associations with viral load. Nat Med 13(1):46–53. doi:10.1038/nm1520

Koup RA, Safrit JT, Cao Y, Andrews CA, McLeod G, Borkowsky W, Farthing C, Ho DD (1994) Temporal association of cellular immune responses with the initial control of viremia in primary human immunodeficiency virus type 1 syndrome. J Virol 68:4650–4655, http://jvi.asm.org/content/68/7/4650.long

Kurihara K, Takahara Y, Nomura T, Ishii H, Iwamoto N, Takahashi N, Inoue M, Iida A, Hara H, Shu T, Hasegawa M, Moriya C, Matano T (2012) Immunogenicity of repeated Sendai viral vector vaccination in macaques. Microbes Infect 14:1169–1176. doi:10.1016/j.micinf.2012.07.016

Kwong PD, Doyle ML, Casper DJ, Cicala C, Leavitt SA, Majeed S, Steenbeke TD, Venturi M, Chaiken I, Fung M, Katinger H, Parren PW, Robinson J, Van Ryk D, Wang L, Burton DR, Freire E, Wyatt R, Sodroski J, Hendrickson WA, Arthos J (2002) HIV-1 evades antibody-mediated neutralization through conformational masking of receptor-binding sites. Nature (Lond) 420:678–682. doi:10.1038/nature01188

Létourneau S, Im EJ, Mashishi T, Brereton C, Bridgeman A, Yang H, Dorrell L, Dong T, Korber B, McMichael AJ, Hanke T (2007) Design and pre-clinical evaluation of a universal HIV-1 vaccine. PLoS One 2(10):e984. doi:10.1371/annotation/fca26a4f-42c1-4772-a19e-aa9d96c4eeb2

Letvin NL, Mascola JR, Sun Y, Gorgone DA, Buzby AP, Xu L, Yang ZY, Chakrabarti B, Rao SS, Schmitz JE, Montefiori DC, Barker BR, Bookstein FL, Nabel GJ (2006) Preserved CD4+ central memory T cells and survival in vaccinated SIV-challenged monkeys. Science 312(5779):1530–1533. doi:10.1126/science.1124226

Li HO, Zhu YF, Asakawa M, Kuma H, Hirata T, Ueda Y, Lee YS, Fukumura M, Iida A, Kato A, Nagai Y, Hasegawa M (2000) A cytoplasmic RNA vector derived from nontransmissible

Sendai virus with efficient gene transfer and expression. J Virol 74(14):6564–6569. doi:10.1128/JVI.74.14.6564-6569.2000

Li Y, Migueles SA, Welcher B, Svehla K, Phogat A, Louder MK, Wu X, Shaw GM, Connors M, Wyatt RT, Mascola JR (2007) Broad HIV-1 neutralization mediated by CD4-binding site antibodies. Nat Med 13(9):1032–1034. doi:10.1038/nm1624

Liu J, O'Brien KL, Lynch DM, Simmons NL, La Porte A, Riggs AM, Abbink P, Coffey RT, Grandpre LE, Seaman MS, Landucci G, Forthal DN, Montefiori DC, Carville A, Mansfield KG, Havenga MJ, Pau MG, Goudsmit J, Barouch DH (2009) Immune control of an SIV challenge by a T-cell-based vaccine in rhesus monkeys. Nature (Lond) 457:87–91. doi:10.1038/nature07469

Lubeck MD, Natuk R, Myagkikh M, Kalyan N, Aldrich K, Sinangil F, Alipanah S, Murthy SC, Chanda PK, Nigida SM Jr, Markham PD, Zolla-Pazner S, Steimer K, Wade M, Reitz MS Jr, Arthur LO, Mizutani S, Davis A, Hung PP, Gallo RC, Eichberg J, Robert-Guroff M (1997) Long-term protection of chimpanzees against high-dose HIV-1 challenge induced by immunization. Nat Med 3:651–658. doi:10.1038/nm0697-651

Machado AV, Cardoso JE, Claser C, Rodrigues MM, Gazzinelli RT, Bruna-Romero O (2006) Long-term protective immunity induced against *Trypanosoma cruzi* infection after vaccination with recombinant adenoviruses encoding amastigote surface protein-2 and trans-sialidase. Hum Gene Ther 17(9):898–908. doi:10.1089/hum.2006.17.898

Mascola JR, Lewis MG, Stiegler G, Harris D, VanCott TC, Hayes D, Louder MK, Brown CR, Sapan CV, Frankel SS, Lu Y, Robb ML, Katinger H, Birx DL (1999) Protection of macaques against pathogenic simian/human immunodeficiency virus 89.6PD by passive transfer of neutralizing antibodies. J Virol 73:4009–4018, http://jvi.asm.org/content/73/5/4009.long

Mascola JR, Stiegler G, VanCott TC, Katinger H, Carpenter CB, Hanson CE, Beary H, Hayes D, Frankel SS, Birx DL, Lewis MG (2000) Protection of macaques against vaginal transmission of a pathogenic HIV-1/SIV chimeric virus by passive infusion of neutralizing antibodies. Nat Med 6:207–210. doi:10.1038/72318

Matano T, Shibata R, Siemon C, Connors M, Lane HC, Martin MA (1998) Administration of an anti-CD8 monoclonal antibody interferes with the clearance of SIV/HIV chimeric virus during primary infections of rhesus macaques. J Virol 72:164–169, http://jvi.asm.org/content/72/1/164.long

Matano T, Kano M, Nakamura H, Takeda A, Nagai Y (2001) Rapid appearance of secondary immune responses and protection from acute CD4 depletion after a highly pathogenic immunodeficiency virus challenge in macaques vaccinated with a DNA-prime/Sendai viral vector-boost regimen. J Virol 75:11891–11896. doi:10.1128/JVI.75.23.11891-11896.2001

Matano T, Kobayashi M, Igarashi H, Takeda A, Nakamura H, Kano M, Sugimoto C, Mori K, Iida A, Hirata T, Hasegawa M, Yuasa T, Miyazawa M, Takahashi Y, Yasunami M, Kimura A, O'Connor DH, Watkins DI, Nagai Y (2004) Cytotoxic T lymphocyte-based control of simian immunodeficiency virus replication in a preclinical AIDS vaccine trial. J Exp Med 199(12):1709–1718. doi:10.1084/jem.20040432

McMichael A (1998) T cell responses and viral escape. Cell 93:673–676. doi:10.1016/S0092-8674(00)81428-2

McShane H, Pathan AA, Sander CR, Keating SM, Gilbert SC, Huygen K, Fletcher HA, Hill AV (2004) Recombinant modified vaccinia virus Ankara expressing antigen 85A boosts BCG-primed and naturally acquired antimycobacterial immunity in humans. Nat Med 10(11):1240–1244. doi:10.1038/nm1128

Menezes H (1968) Protective effect of an avirulent (cultivated) strain of *Trypanosoma cruzi* against experimental infection in mice. Rev Inst Med Trop Sao Paulo 10(1):1–4

Moore JP, Cao Y, Qing L, Sattentau QJ, Pyati J, Koduri R, Robinson J, Barbas CF 3rd, Burton DR, Ho DD (1995) Primary isolates of human immunodeficiency virus type 1 are relatively resistant to neutralization by monoclonal antibodies to gp120, and their neutralization is not predicted by studies with monomeric gp120. J Virol 69:101–109, http://jvi.asm.org/content/69/1/101.long

Moore JP, Parren PW, Burton DR (2001) Genetic subtypes, humoral immunity, and human immunodeficiency virus type 1 vaccine development. J Virol 75:5721–5729. doi:10.1128/JVI.75.13.5721-5729.2001

Moriya C, Horiba S, Inoue M, Iida A, Hara H, Shu T, Hasegawa M, Matano T (2008) Antigen-specific T-cell induction by vaccination with a recombinant Sendai virus vector even in the presence of vector-specific neutralizing antibodies in rhesus macaques. Biochem Biophys Res Commun 371(4):850–854. doi:10.1016/j.bbrc.2008.04.156

Moriya C, Horiba S, Kurihara K, Kamada T, Takahara Y, Inoue M, Iida A, Hara H, Shu T, Hasegawa M, Matano T (2011) Intranasal Sendai viral vector vaccination is more immunogenic than intramuscular under pre-existing anti-vector antibodies. Vaccine 29(47):8557–8563. doi:10.1016/j.vaccine.2011.09.028

Mossman SP, Bex F, Berglund P, Arthos J, O'Neil SP, Riley D, Maul DH, Bruck C, Momin P, Burny A, Fultz PN, Mullins JI, Liljeström P, Hoover EA (1996) Protection against lethal simian immunodeficiency virus SIVsmmPBj14 disease by a recombinant Semliki Forest virus gp160 vaccine and by a gp120 subunit vaccine. J Virol 70:1953–1960, http://jvi.asm.org/content/70/3/1953.long

Myszka DG, Sweet RW, Hensley P, Brigham-Burke M, Kwong PD, Hendrickson WA, Wyatt R, Sodroski J, Doyle ML (2000) Energetics of the HIV gp120-CD4 binding reaction. Proc Natl Acad Sci USA 97(16):9026–9031. doi:10.1073/pnas.97.16.9026

Naruse TK, Chen Z, Yanagida R, Yamashita T, Saito Y, Mori K, Akari H, Yasutomi Y, Miyazawa M, Matano T, Kimura A (2010) Diversity of MHC class I genes in Burmese-origin rhesus macaques. Immunogenetics 62:601–611. doi:10.1007/s00251-010-0462-z

Nishimura Y, Igarashi T, Haigwood NL, Sadjadpour R, Donau OK, Buckler C, Plishka RJ, Buckler-White A, Martin MA (2003) Transfer of neutralizing IgG to macaques 6 h but not 24 h after SHIV infection confers sterilizing protection: implications for HIV-1 vaccine development. Proc Natl Acad Sci USA 100:15131–15136. doi:10.1073/pnas.2436476100

Nomura T, Yamamoto H, Shiino T, Takahashi N, Nakane T, Iwamoto N, Ishii H, Tsukamoto T, Kawada M, Matsuoka S, Takeda A, Terahara K, Tsunetsugu-Yokota Y, Iwata-Yoshikawa N, Hasegawa H, Sata T, Naruse TK, Kimura A, Matano T (2012) Association of major histocompatibility complex class I haplotypes with disease progression after simian immunodeficiency virus challenge in Burmese rhesus macaques. J Virol 86:6481–6490. doi:10.1128/JVI.07077-11

Parren PW, Marx PA, Hessell AJ, Luckay A, Harouse J, Cheng-Mayer C, Moore JP, Burton DR (2001) Antibody protects macaques against vaginal challenge with a pathogenic R5 simian/human immunodeficiency virus at serum levels giving complete neutralization in vitro. J Virol 75:8340–8347. doi:10.1128/JVI.75.17.8340-8347.2001

Pitisuttithum P, Gilbert P, Gurwith M, Heyward W, Martin M, van Griensven F, Hu D, Tappero JW, Choopanya K, Bangkok Vaccine Evaluation Group (2006) Randomized, double-blind, placebo-controlled efficacy trial of a bivalent recombinant glycoprotein 120 HIV-1 vaccine among injection drug users in Bangkok, Thailand. J Infect Dis 194(12):1661–1671. doi:10.1086/508748

Reitter JN, Means RE, Desrosiers RC (1998) A role for carbohydrates in immune evasion in AIDS. Nat Med 4:679–684. doi:10.1038/nm0698-679

Rerks-Ngarm S, Pitisuttithum P, Nitayaphan S, Kaewkungwal J, Chiu J, Paris R, Premsri N, Namwat C, de Souza M, Adams E, Benenson M, Gurunathan S, Tartaglia J, McNeil JG, Francis DP, Stablein D, Birx DL, Chunsuttiwat S, Khamboonruang C, Thongcharoen P, Robb ML, Michael NL, Kunasol P, Kim JH, MOPH-TAVEG Investigators (2009) Vaccination with ALVAC and AIDSVAX to prevent HIV-1 infection in Thailand. N Engl J Med 361(23):2209–2220, http://www.nejm.org/doi/full/10.1056/NEJMoa0908492

Richman DD, Wrin T, Little SJ, Petropoulos CJ (2003) Rapid evolution of the neutralizing antibody response to HIV type 1 infection. Proc Natl Acad Sci USA 100:4144–4149. doi:10.1073/pnas.0630530100

Rolland M, Nickle DC, Mullins JI (2007) HIV-1 group M conserved elements vaccine. PLoS Pathog 3(11):e157. doi:10.1371/journal.ppat.0030157

Rolland M, Tovanabutra S, deCamp AC, Frahm N, Gilbert PB, Sanders-Buell E, Heath L, Magaret CA, Bose M, Bradfield A, O'Sullivan A, Crossler J, Jones T, Nau M, Wong K, Zhao H, Raugi DN, Sorensen S, Stoddard JN, Maust BS, Deng W, Hural J, Dubey S, Michael NL, Shiver J, Corey L, Li F, Self SG, Kim J, Buchbinder S, Casimiro DR, Robertson MN, Duerr A, McElrath

MJ, McCutchan FE, Mullins JI (2011) Genetic impact of vaccination on breakthrough HIV-1 sequences from the STEP trial. Nat Med 17(3):366–371. doi:10.1038/nm.2316

Rose NF, Marx PA, Luckay A, Nixon DF, Moretto WJ, Donahoe SM, Montefiori D, Roberts A, Buonocore L, Rose JK (2001) An effective AIDS vaccine based on live attenuated vesicular stomatitis virus recombinants. Cell 106:539–549. doi:10.1016/S0092-8674(01)00482-2

Santra S, Liao HX, Zhang R, Muldoon M, Watson S, Fischer W, Theiler J, Szinger J, Balachandran H, Buzby A, Quinn D, Parks RJ, Tsao CY, Carville A, Mansfield KG, Pavlakis GN, Felber BK, Haynes BF, Korber BT, Letvin NL (2010) Mosaic vaccines elicit CD8+ T lymphocyte responses that confer enhanced immune coverage of diverse HIV strains in monkeys. Nat Med 16(3):324–328. doi:10.1038/nm.2108

Scheid JF, Mouquet H, Ueberheide B, Diskin R, Klein F, Oliveira TY, Pietzsch J, Fenyo D, Abadir A, Velinzon K, Hurley A, Myung S, Boulad F, Poignard P, Burton DR, Pereyra F, Ho DD, Walker BD, Seaman MS, Bjorkman PJ, Chait BT, Nussenzweig MC (2011) Sequence and structural convergence of broad and potent HIV antibodies that mimic CD4 binding. Science 333(6049):1633–1637. doi:10.1126/science.1207227

Shibata R, Igarashi T, Haigwood N, Buckler-White A, Ogert R, Ross W, Willey R, Cho MW, Martin MA (1999) Neutralizing antibody directed against the HIV-1 envelope glycoprotein can completely block HIV-1/SIV chimeric virus infections of macaque monkeys. Nat Med 5:204–210. doi:10.1038/5568

Shiver JW, Fu TM, Chen L, Casimiro DR, Davies ME, Evans RK, Zhang ZQ, Simon AJ, Trigona WL, Dubey SA, Huang L, Harris VA, Long RS, Liang X, Handt L, Schleif WA, Zhu L, Freed DC, Persaud NV, Guan L, Punt KS, Tang A, Chen M, Wilson KA, Collins KB, Heidecker GJ, Fernandez VR, Perry HC, Joyce JG, Grimm KM, Cook JC, Keller PM, Kresock DS, Mach H, Troutman RD, Isopi LA, Williams DM, Xu Z, Bohannon KE, Volkin DB, Montefiori DC, Miura A, Krivulka GR, Lifton MA, Kuroda MJ, Schmitz JE, Letvin NL, Caulfield MJ, Bett AJ, Youil R, Kaslow DC, Emini EA (2002) Replication-incompetent adenoviral vaccine vector elicits effective anti-immunodeficiency-virus immunity. Nature (Lond) 415:331–335. doi:10.1038/415331a

Slobod KS, Shenep JL, Luján-Zilbermann J, Allison K, Brown B, Scroggs RA, Portner A, Coleclough C, Hurwitz JL (2004) Safety and immunogenicity of intranasal murine parainfluenza virus type 1 (Sendai virus) in healthy human adults. Vaccine 22(23–24):3182–3186. doi:10.1016/j.vaccine.2004.01.053

Stamatatos L, Morris L, Burton DR, Mascola JR (2009) Neutralizing antibodies generated during natural HIV-1 infection: good news for an HIV-1 vaccine? Nat Med 15(8):866–870. doi:10.1038/nm.1949

Takeda A, Igarashi H, Nakamura H, Kano M, Iida A, Hirata T, Hasegawa M, Nagai Y, Matano T (2003) Protective efficacy of an AIDS vaccine, a single DNA-prime followed by a single booster with a recombinant replication-defective Sendai virus vector, in a macaque AIDS model. J Virol 77:9710–9715. doi:10.1128/JVI.77.17.9710-9715.2003

Takeda A, Igarashi H, Kawada M, Tsukamoto T, Yamamoto H, Inoue M, Iida A, Shu T, Hasegawa M, Matano T (2008) Evaluation of the immunogenicity of replication-competent V-knocked-out and replication-defective F-deleted Sendai virus vector-based vaccines in macaques. Vaccine 26:6839–6843. doi:10.1016/j.vaccine.2008.09.074

Townsend AR, Rothbard J, Gotch FM, Bahadur G, Wraith D, McMichael AJ (1986) The epitopes of influenza nucleoprotein recognized by cytotoxic T lymphocytes can be defined with short synthetic peptides. Cell 44:959–968. doi:10.1016/0092-8674(86)90019-X

Trkola A, Kuster H, Rusert P, Joos B, Fischer M, Leemann C, Manrique A, Huber M, Rehr M, Oxenius A, Weber R, Stiegler G, Vcelar B, Katinger H, Aceto L, Günthard HF (2005) Delay of HIV-1 rebound after cessation of antiretroviral therapy through passive transfer of human neutralizing antibodies. Nat Med 11:615–622. doi:10.1038/nm1244

Tsukamoto T, Takeda A, Yamamoto T, Yamamoto H, Kawada M, Matano T (2009) Impact of cytotoxic-T-lymphocyte memory induction without virus-specific CD4+ T-cell help on control of a simian immunodeficiency virus challenge in rhesus macaques. J Virol 83:9339–9346

Vasconcelos JR, Hiyane MI, Marinho CR, Claser C, Machado AM, Gazzinelli RT, Bruña-Romero O, Alvarez JM, Boscardin SB, Rodrigues MM (2004) Protective immunity against *Trypanosoma cruzi* infection in a highly susceptible mouse strain after vaccination with genes encoding the amastigote surface protein-2 and trans-sialidase. Hum Gene Ther 15(9):878–886. doi:10.1089/hum.2004.15.878

Veazey RS, Shattock RJ, Pope M, Kirijan JC, Jones J, Hu Q, Ketas T, Marx PA, Klasse PJ, Burton DR, Moore JP (2003) Prevention of virus transmission to macaque monkeys by a vaginally applied monoclonal antibody to HIV-1 gp120. Nat Med 9:343–346. doi:10.1038/nm833

Walker LM, Simek MD, Priddy F, Gach JS, Wagner D, Zwick MB, Phogat SK, Poignard P, Burton DR (2010) A limited number of antibody specificities mediate broad and potent serum neutralization in selected HIV-1 infected individuals. PLoS Pathog 6(8):e1001028. doi:10.1371/journal.ppat.1001028

Walker LM, Huber M, Doores KJ, Falkowska E, Pejchal R, Julien JP, Wang SK, Ramos A, Chan-Hui PY, Moyle M, Mitcham JL, Hammond PW, Olsen OA, Phung P, Fling S, Wong CH, Phogat S, Wrin T, Simek MD, Koff WC, Wilson IA, Burton DR, Poignard P (2011) Broad neutralization coverage of HIV by multiple highly potent antibodies. Nature (Lond) 477(7365):466–470. doi:10.1038/nature10373

Webster DP, Dunachie S, McConkey S, Poulton I, Moore AC, Walther M, Laidlaw SM, Peto T, Skinner MA, Gilbert SC, Hill AV (2006) Safety of recombinant fowlpox strain FP9 and modified vaccinia virus Ankara vaccines against liver-stage *P. falciparum* malaria in non-immune volunteers. Vaccine 24(15):3026–3034. doi:10.1016/j.vaccine.2005.10.058

Wei X, Decker JM, Wang S, Hui H, Kappes JC, Wu X, Salazar-Gonzalez JF, Salazar MG, Kilby JM, Saag MS, Komarova NL, Nowak MA, Hahn BH, Kwong PD, Shaw GM (2003) Antibody neutralization and escape by HIV-1. Nature (Lond) 422:307–312. doi:10.1038/nature01470

Wilson NA, Reed J, Napoe GS, Piaskowski S, Szymanski A, Furlott J, Gonzalez EJ, Yant LJ, Maness NJ, May GE, Soma T, Reynolds MR, Rakasz E, Rudersdorf R, McDermott AB, O'Connor DH, Friedrich TC, Allison DB, Patki A, Picker LJ, Burton DR, Lin J, Huang L, Patel D, Heindecker G, Fan J, Citron M, Horton M, Wang F, Liang X, Shiver JW, Casimiro DR, Watkins DI (2006) Vaccine-induced cellular immune responses reduce plasma viral concentrations after repeated low-dose challenge with pathogenic simian immunodeficiency virus SIVmac239. J Virol 80(12):5875–5885. doi:10.1128/JVI.00171-06

Wizel B, Garg N, Tarleton RL (1998) Vaccination with trypomastigote surface antigen 1-encoding plasmid DNA confers protection against lethal *Trypanosoma cruzi* infection. Infect Immun 66(11):5073–5081, http://iai.asm.org/content/66/11/5073.long

Wu X, Zhou T, Zhu J, Zhang B, Georgiev I, Wang C, Chen X, Longo NS, Louder M, McKee K, O'Dell S, Perfetto S, Schmidt SD, Shi W, Wu L, Yang Y, Yang ZY, Yang Z, Zhang Z, Bonsignori M, Crump JA, Kapiga SH, Sam NE, Haynes BF, Simek M, Burton DR, Koff WC, Doria-Rose NA, Connors M, NISC Comparative Sequencing Program, Mullikin JC, Nabel GJ, Roederer M, Shapiro L, Kwong PD, Mascola JR (2011) Focused evolution of HIV-1 neutralizing antibodies revealed by structures and deep sequencing. Science 333(6049):1593–1602. doi:10.1126/science.1207532

Wyatt R, Kwong PD, Desjardins E, Sweet RW, Robinson J, Hendrickson WA, Sodroski JG (1998) The antigenic structure of the HIV gp120 envelope glycoprotein. Nature (Lond) 393:705–711. doi:10.1038/31514

Xin KQ, Urabe M, Yang J, Nomiyama K, Mizukami H, Hamajima K, Nomiyama H, Saito T, Imai M, Monahan J, Okuda K, Ozawa K, Okuda K (2001) A novel recombinant adeno-associated virus vaccine induces a long-term humoral immune response to human immunodeficiency virus. Hum Gene Ther 12(9):1047–1061. doi:10.1089/104303401750214276

Yamamoto H, Matano T (2010) Neutralizing antibodies in SIV control: co-impact with T cells. Vaccine 28S:B13–B17. doi:10.1016/j.vaccine.2009.09.080

Yamamoto H, Kawada M, Takeda A, Igarashi H, Matano T (2007) Post-infection immunodeficiency virus control by neutralizing antibodies. PLoS One 2(6):e540. doi:10.1371/journal.pone.0000540

Yamamoto T, Iwamoto N, Yamamoto H, Tsukamoto T, Kuwano T, Takeda A, Kawada M, Tsunetsugu-Yokota Y, Matano T (2009) Polyfunctional $CD4^+$ T-cell induction in neutralizing antibody-triggered control of simian immunodeficiency virus infection. J Virol 83:5514–5524. doi:10.1128/JVI.00145-09

Zhou T et al (2010) Structural basis for broad and potent neutralization of HIV-1 by antibody VRC01. Science 329:811–817. doi:10.1126/science.1192819

Zuñiga R, Lucchetti A, Galvan P, Sanchez S, Sanchez C, Hernandez A, Sanchez H, Frahm N, Linde CH, Hewitt HS, Hildebrand W, Altfeld M, Allen TM, Walker BD, Korber BT, Leitner T, Sanchez J, Brander C (2006) Relative dominance of Gag p24-specific cytotoxic T lymphocytes is associated with human immunodeficiency virus control. J Virol 80(6):3122–3125. doi:10.1128/JVI.80.6.3122-3125.2006

Chapter 6
BioKnife, a Modified Sendai Virus, to Resect Malignant Tumors

Yoshikazu Yonemitsu, Yasuji Ueda, and Mamoru Hasegawa

Abstract The *M* gene-deleted SeV, **Δ***M*SeV, was found to induce massive cell–cell fusion (syncytia formation) leading to extensive cell death of the entire monolayer culture upon cleavage activation of the viral F glycoprotein on the infected cell surface (see Chap. 3). This finding suggested that **Δ***M*SeV would have great potential as an oncolytic agent for a solid malignant tumor if it could be further engineered to selectively target the tumor cells. For this targeting we made use of the theory of protease-dependent tissue tropism of SeV (see Chap. 2). Namely, we attempted to render the inactive precursor F_0 protein cleavage site of **Δ***M*SeV susceptible to the proteases overexpressed by diverse tumor cells, including the matrix metalloproteases (MMPs) and urokinase-type plasminogen activator (uPA). In addition, a portion of the cytoplasmic tail of the F protein was deleted to maximize the fusion-inducing capacity. The resultant MMP-targeted and uPA-targeted **Δ***M*SeVs displayed highly tumor cell-specific killing at the cellular and animal levels. Of these, a uPA-targeted **Δ***M*SeV appeared to be particularly useful because of its remarkably high efficacy in eradication of tumors in various animal models in addition to its potential diversity of therapeutic targets. We named this virus "BioKnife" because of its desirable nature to resect diverse malignant solid tumors without damaging the surrounding healthy tissues in preclinical studies using animal models. We thus propose a conceptually new strategy in designing an oncolytic virus. Further studies are eagerly awaited to assess its safety and efficacy in clinical settings.

Y. Yonemitsu
Faculty of Pharmaceutical Sciences, Kyushu University, 3-1-1, Maidashi, Higashi-ku, Fukuoka-shi, Fukuoka 812-8582, Japan

Y. Ueda • M. Hasegawa (✉)
DNAVEC Corporation, 6 Ohkubo, Tsukuba-shi, Ibaraki 300-2611, Japan
e-mail: hasegawa@dnavec-corp.com

Y. Nagai (ed.), *Sendai Virus Vector: Advantages and Applications*,
DOI 10.1007/978-4-431-54556-9_6, © Springer Japan 2013

6.1 Introduction

6.1.1 Background

Oncolytic virotherapy is a concept to selectively kill malignant tumor cells by viruses. In the 1950s and 1960s there were numerous attempts to direct the evolution of animal viruses toward greater tumor specificity, but success was limited and many researchers abandoned the idea. However, the subsequent development of recombinant DNA technology and more recent reverse genetics for RNA viruses reignited interest in virotherapy, allowing the generation of more potent tumor-specific oncolytic virus variants. Oncolytic virotherapy can thus be considered a reemerging field, rather than an emerging one (reviewed in Kelly and Russell 2007). Diverse virus species are now under active investigation, and some of them have left the laboratory for clinical trials (reviewed in Russell et al. 2012).

In these reemerging studies, two factors are crucially important: the viral abilities to target tumor cells selectively and to kill them effectively. To this end, diverse attempts have been made, as exemplified next.

Newcastle disease virus (NDV) and vesicular stomatitis virus (VSV), which belong to the families *Paramyxoviridae* and *Rhabdoviridae*, respectively, have been used because of their natural preference to tumor cells (Russell et al. 2012). Adenoviruses have been used that were engineered for their replication to be telomerase dependent (Fujiwara et al. 2008). The adenovirus attachment protein was engineered to recognize tumor-specific cell-surface molecules, conferring higher tumor selectivity on the viruses (Chiocca 2002; Ring 2002; Russell et al. 2012). Herpesvirus accessory genes are indispensable for the viral replication in normal cells but dispensable in tumor cells, leading to generation of antitumor herpesviruses by deleting the accessory genes (Todo 2012). The approach further has included the use of various virus species encoding suicide genes activating pro-antitumor drugs, expecting the killing of not only infected cells but also the surrounding tumor cells (bystander effect).

6.1.2 Toward the BioKnife by Manipulating Sendai Virus (SeV)

As described in Chap. 3, we have expanded the plasmid-based reverse genetics technology originally developed for the manipulation of the wild-type SeV genome to delete certain viral genes, primarily to generate nontransmissible, safer versions of SeV vectors for gene therapy and gene vaccination. Of these versions the *M* (matrix protein) gene-deleted SeV (Δ*M*SeV, Fig. 6.1a) is unique in that it does not bud to form mature viral particles from the surface of an infected cell but instead rapidly spreads from cell to cell in the presence of trypsin activating the F (fusion) glycoprotein. Here, massive cell–cell fusion (syncytium formation) was induced in the entire monolayer culture of cells, followed by their rapid and complete death

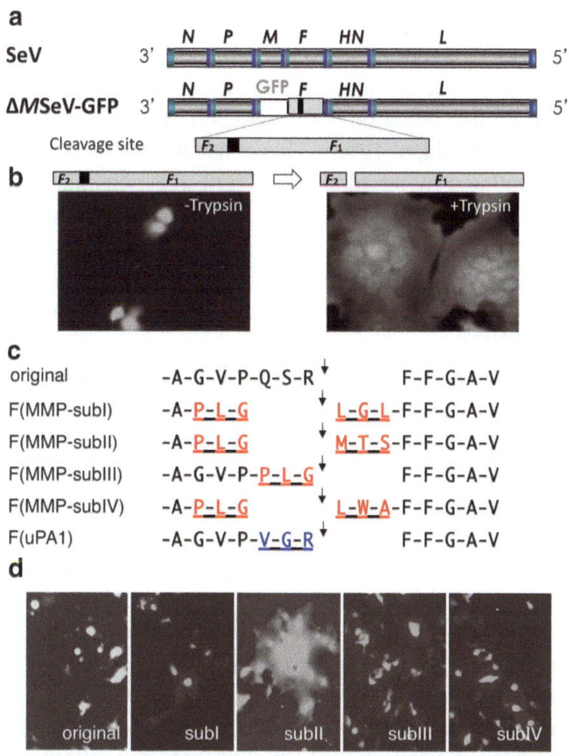

Fig. 6.1 (**a**) Gene maps of the wild-type Sendai virus (SeV) and *M* gene-deleted SeV (Δ*M*SeV). In Δ*M*SeV, the ORF of the M protein was replaced by the GFP ORF. (**b**) Trypsin-dependent syncytial formation with transfection by Δ*M*SeV. (**c**) Amino acid changes at the F-protein cleavage site. The amino acid sequence was changed to that of a matrix metalloprotease (MMP) substrate (MMP-subI–subIV) or a urokinase-type plasminogen activator (uPA) substrate (uPA1). (**d**) Capacity of the resulting F-protein mutants to induce cell–cell fusion (syncytium formation) analyzed by transfection of MMP-expressing HT1080 cells with the *F* gene encoding plasmid with modified cleavage sequences, together with the *HN* gene-encoding plasmid. (Adapted from Kinoh et al. 2009)

(Inoue et al. 2003; Kinoh et al. 2004; Fig. 6.1b). Thus, Δ*M*SeV appeared to fulfill one of the two requirements of solid tumor-lytic viruses, the strong cell-killing ability; that is, multiple rounds of cell-to-cell spreading from a limited number of infected cells to the entire tumor, ultimately wiping out every cancer cell in a solid tumor. We used the Δ*M*SeV as the starting material and attempted to achieve the other requirement, tumor-selective infection, by applying the concept of "protease-dependent viral tropism" of SeV (see Chap. 2).

SeV displays a narrow spectrum of tissue tropism in susceptible hosts, replicating in the respiratory tracts of mice or in the allantois of embryonated chicken eggs with little appreciable spreading to other tissues, even though its receptors are terminal sialic acid residues, which are ubiquitous throughout the animal body. This restricted tropism is primarily because specific tissue proteases required for cleavage activation of the viral fusion (F) glycoprotein from the inactive precursor F_0 to the active F_1 and F_2 subunits (Fig. 6.1a), and thus for the infectivity of progeny (the capacity to penetrate into and initiate infection in the next cell), are expressed on the surface of limited types of tissues (Nagai 1993). This concept of protease-dependent tropism of SeV was substantiated by the identification of a responsible tissue protease in a model organism, the chick embryo (see Chap. 2). The identified protease was the blood clotting factor Xa ectopically expressed by allantoic or amniotic cells and secreted into the allantoic or amniotic fluid (Gotoh et al. 1990). The spreading in ovo was limited to those tissues in direct contact with the fluid containing FXa (Ogasawara et al. 1992).

It has to be noted that the cleavage (↓) motif Q-S-R↓ is shared with the natural substrate prothrombin of FXa and the SeV F_0 protein (Fig. 6.1c). Most of the cells in culture and the tissues in ovo tested so far did not express SeV-activating proteases and hence have to be supplemented with a low concentration of exogenously added trypsin, a less-specific digestive enzyme, to allow SeV spreading (Homma and Ouchi 1973).

The extracellular matrix (ECM) surrounding a tumor cell serves as a barrier that blocks rapid tumor cell division, migration, infiltration, and metastasis. Highly invasive, metastatic tumor cells express high levels of ECM-degrading enzymes such as matrix metalloproteases (MMPs) and urokinase-type plasminogen activator (uPA) (reviewed in Andreasen et al. 2000; Cox and O'Byrne 2001).

Taking into consideration the protease-dependent SeV tropism and the high level of ECM-degrading enzyme expression by malignant tumor cells together, we initiated extensive efforts to learn whether the natural cleavage/activation can be converted to tumor cell-selective activation by manipulating the *F* gene in Δ*M*SeV.

Indeed, the efforts were successful, generating a series of MMP- or uPA-activated Δ*M*SeV and facilitating the proof-of-concept studies *in vitro* using cell cultures and *in vivo* using animal models. Overall, one of the uPA-targeted Δ*M*SeV constructs looked like a sharp biological knife used to resect various malignant tumors without injuring surrounding tissues and thus was named "BioKnife".

6.2 Designing Tumor-Targeting/Lysing SeV: Proof-of-Concept Studies *In Vitro*

6.2.1 Massive Syncytia Formation by the M-Gene-Deleted SeV (ΔMSeV): A Hint for the Development of Oncolytic SeV

When LLCMK2 cells were infected with Δ*M*SeV, even at an MOI lower than 0.1, and incubated in the presence of a small amount of trypsin, massive syncytia rapidly developed in the entire monolayer (Fig. 6.1b); this was followed by complete death of the syncytia. It appeared that the RNPs of Δ*M*SeV accumulated in cells spread by this cell–cell fusion to the entire syncytium and there express viral proteins including the F and HN proteins, accelerating the next round of cell–cell fusion on one hand and on the other the death of the syncytia. Such extensive cell to cell fusion followed by rapid cell killing hinted to us that this mutant would become a potential tool for lysing solid tumors, if it would be further engineered to target the tumor tissues.

6.2.2 *Plasmid-Based Studies*

The cleavage of the SeV precursor F_0 protein to F_1 and F_2 takes place between the R and F residues at positions 116 and 117 of the F_0 sequence, respectively (Fig. 6.1c; see also Chap. 2). There are a number of known MMPs. According to the results obtained using synthetic substrates for the MMP assay, the cleavage sites of their substrates appeared to consist of the tripeptide PLG, followed by the other tripeptides such as LGL (see MMP-sub I in Fig. 6.1c; Kinoh et al. 2004, and references therein). Cleavage is expected to take place between the upstream and downstream tripeptides. The counterpart tripeptides in the SeV F_0 protein are QSR and FFG, respectively, which are largely different from those of the MMP substrates.

6.2.2.1 Introduction of MMP Cleavage Sites into the F_0 Protein

The N-terminal hydrophobic FFG of SeV F_1 is followed by an additional 23 hydrophobic residues. The stretch of 26 hydrophobic amino acids (the fusion peptide) is thought to be inserted into the target cell membrane and is thus a critical component required for the cell–virus membrane fusion process. This early event in the fusion process requires the coexpression of the receptor-binding HN protein (see Chap. 2). It was shown that the carbobenzoxy-FFG oligopeptide could specifically inhibit SeV fusion activity and infectivity in cultured cells, suggesting that the N-terminal FFG was specifically important in the fusion process (Richardson et al. 1980). We therefore wondered whether we would be able to introduce MMP-cleavable sites into the F_0 protein and whether such newly designed F_0 proteins could be activated by MMPs to become a fusion-competent form.

In a preliminary study using the plasmid encoding the SeV F protein, we changed its natural cleavage site sequence to each of the four sites potentially cleavable by MMPs (MMP-subI–subIV; Fig. 6.1c) and assayed their syncytium formation activity after co-transfection with an HN-expressing plasmid. We found that F (MMP-subII) protein could induce potent syncytia in the HT1080 human fibrosarcoma cell line, which is known to express the specific MMP (Kinoh et al. 2004; subII in Fig. 6.1d). The newly generated fusion-competent F_1 was expected to contain MTS at its N terminus, which was rather surprising in view of the foregoing studies suggesting the importance of the exposure of the terminal FFG after cleavage (Richardson et al. 1980). In addition, MTS is hydrophilic and could disturb the access and insertion into the target cell membrane of the fusion peptide. Nevertheless, cell fusion actually occurred, indicating that the F_1 terminal MTS was not always obstructive. Alternatively, the MTS might be removed by amino peptidase(s) during

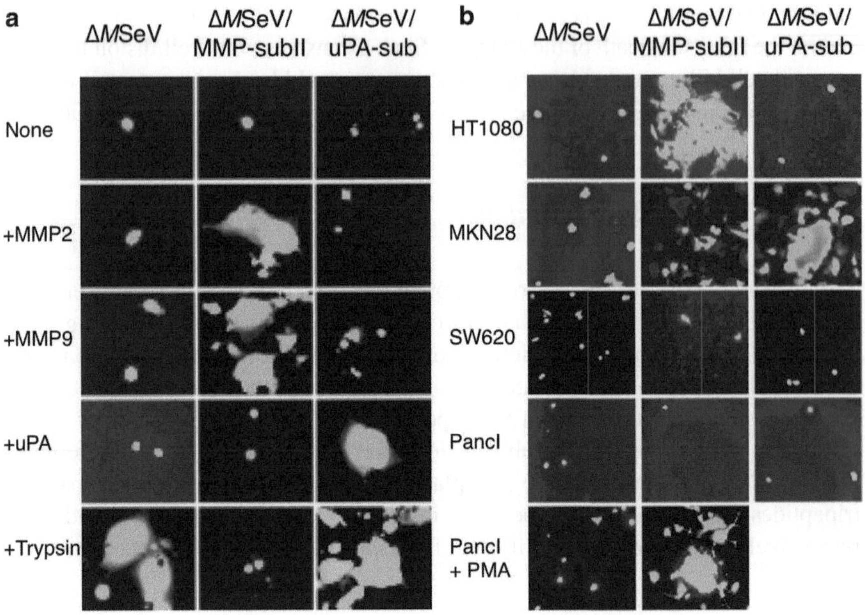

Fig. 6.2 (**a**) Alterations of protease requirement and cell tropism of Δ*M*SeV with modified cleavage sites of F glycoprotein. LLC-MK2 cells infected with the parental Δ*M*SeV, Δ*M*SeV/MMP-subII, or Δ*M*SeV/uPA-sub were cultured with exogenously added MMP2, MMP9, uPA, or trypsin. (**b**) Alterations of protease requirement and cell tropism of Δ*M*SeV shown by endogenous proteases. Four tumor cell lines, HT1080, MKN28, SW620, and Panc I, were infected with the parental Δ*M* SeV, Δ*M*SeV/MMP-subII, or Δ*M*SeV/uPA-sub. Panc I cells were also analyzed to assess syncytia formation in the presence of MMP induced by the addition of phorbol 12-myristate 13-acetate (PMA). (Adapted from Kinoh et al. 2004)

or after cell-surface expression of the F glycoprotein. These issues need to be addressed in future studies of SeV biology.

The three other potential cleavage motifs, designated as subI, subIII, and subIV (Fig. 6.1c), failed to induce syncytia of HT1080. It remains to be explained whether this failure is the result of poor cleavability of the designed F_0 or the addition of LGL (subI) or LWA (subIV) to the F_1 N-terminus.

6.2.2.2 Introduction of uPA Cleavage Site into the F_0 Protein

uPA and SeV-activating proteases (e.g., FXa) are trypsin-like serine proteases and cleave at the carboxyl side of R residues at their cleavage sites. They share the same upstream sequence, AGVP (Fig. 6.1c). We succeeded in generating a new F glycoprotein that was activated by uPA as well as by trypsin (data not shown; see Fig. 6.2 for reference) by changing two amino acids, QS to VG, in the upstream region of the cleavage site (Fig. 6.1c).

Overall, these plasmid-based studies strongly suggested that the creation of Δ*M*SeV harboring F_0 selectively cleavable by at least some MMP species or uPA would be possible.

6.2.3 *Recovery of* Δ*MSeV Harboring the* F_0 *Proteins Susceptible to MMPs or uPA*

6.2.3.1 MMP-Susceptible Δ*M*SeV

The Δ*M*SeV with MMP-subII sequence in its F_0 protein (Δ*M*SeV/MMP-subII) (Fig. 6.1c) was successfully recovered (for the method of recovery, see Chap. 3) and found to be able to induce remarkable syncytia formation in LLCMK2 cells in the presence of MMP2 or MMP9, but not at all in the presence of trypsin (Kinoh et al. 2004; Fig. 6.2a). The parental Δ*M*SeV that served as a control exhibited syncytia formation by trypsin but not by MMP2 or MMP9. It should be noted that MMP2 and MMP9, together with MMP7, are likely to be more important for tumor metastasis than the 20 other known MMP species because only the former 3 have the capacity to disrupt basement membranes (Friedberg et al. 1998; Curran and Murray 1999).

6.2.3.2 uPA-Susceptible Δ*M*SeV

The Δ*M*SeV harboring the uPA cleavage site in its F_0 protein (Δ*M*SeV/uPA-sub) was also recovered. The virus was found to induce syncytia in LLCMK2 cells when supplemented with uPA as well as trypsin, but not with MMP2 or MMP9 (Fig. 6.2a). uPA is well known to be an essential factor produced by a variety of malignant tumor cells for their rapid cell division, migration, and metastasis (Andreasen et al. 2000; Cox and O'Byrne 2001), and thus the virus was expected to represent a powerful tool to suppress the virulence of such tumors.

6.2.3.3 *In Vitro* Studies Using Tumor Cell Lines Expressing MMPs or uPA

In perfect correlation with the foregoing results obtained with exogenously added proteases, Δ*M*SeV/MMP-subII induced extensive syncytia formation in a human fibrosarcoma cell line, HT1080, endogenously secreting MMPs, but not in a human stomach cancer line, MKN28, secreting uPA but not MMPs. On the other hand, Δ*M*SeV/uPA-sub showed the opposite result, forming syncytia in MKN28 but not in HT1080 (Fig. 6.2b). Neither of the cell lines showed syncytia formation with the parental Δ*M*SeV.

A human pancreatic epithelioid carcinoma cell line, Panc I, expresses MMP2 and MMP9 at lower levels than HT1080. This cell line was unable to activate

Δ*M*SeV/MMP-subII (Fig. 6.2b). However, Panc I cells developed syncytia with the modified Δ*M*SeV when MMP9 production was enhanced by treating the cells with a phorbol ester, phorbol 12-myristate 13-acetate (PMA) (Fig. 6.2b). In this experiment, SW620 (a human lymph node colorectal adenocarcinoma cell line) was used as a negative control to show no syncytia formation because of the small production of MMPs or uPA by the cell line.

Taken together, these *in vitro* studies appeared to validate our concept of creating malignant tumor-targeting SeV by making use of the protease-dependent theory of SeV tropism. From the opposite point of view, the current results reinforce the protease-dependent theory of the cell/tissue tropism of SeV and related viruses.

6.3 Optimization of Δ*M*SeV to Confer the Highest Fusogenic and Oncolytic Activity

Before moving on to *in vivo* animal studies, we used two different approaches to confer higher oncolytic potential on Δ*M*SeV, namely, the manipulation of F glycoprotein at its cytoplasmic domain and cleavage-site sequences targeted by MMPs and uPA.

6.3.1 Enhancement of the Fusogenic Activity of F Glycoprotein by Deletion of Its Cytoplasmic Tail

Earlier studies demonstrated that the truncation of cytoplasmic domains in the glycoproteins of not only retrovirus and herpesvirus (Rein et al. 1994), but also of *Paramyxoviridae*, that is, measles virus and Newcastle disease virus, increased their fusogenic activity (Cathomen et al. 1998). We examined the effect on SeV fusion activity of the deletions of three different lengths: 27, 14, and 4 residues from the C-terminus in the cytoplasmic tail of the F glycoprotein. The plasmids encoding these truncation mutant proteins were named Fct27, Fct14, and Fct4, respectively. Each of them was co-transduced with the HN protein-expressing plasmid into LLCMK2 cells, and the resulting syncytium-inducing activities of these truncated F proteins (in the presence of trypsin) were compared with that of the wild-type F glycoprotein. The activity was measured by counting the number of nuclei in the syncytia formed through cell–cell fusion. Fct14 was found to be far greater in terms of syncytium-forming capacity than the other two truncations and the wild-type F glycoproteins (Kinoh et al. 2009). The recombinant virus with that deletion was named Δ*M*SeV/Fct14 and was preferentially used in the subsequent studies.

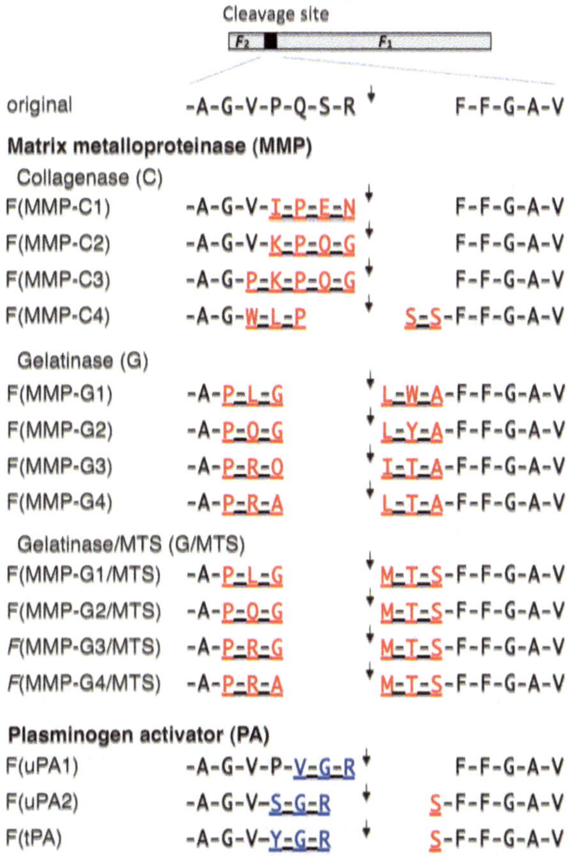

Fig. 6.3 Optimization of linker sequences susceptible to MMP and uPA. Designated substrate sequences for collagenase (MMP1, MMP3, MT-MMP, etc.), gelatinase (MMP2, MMP9), and plasminogen activator (uPA, tPA) are shown. Effects of additional MTS sequences were also examined (MMP-G1 to -GM4/MTS). For details, see text. *tPA*, tissue plasminogen activator. (Adapted from Kinoh et al. 2009)

6.3.2 *Optimization of the F_0 Cleavage Site Sequence*

There are numerous other sequences besides those shown in Fig. 6.1c that can presumably be placed at the cleavage site of SeV F_0 protein (for details, see Kinoh et al. 2009). We designed a total of 12 candidate sequences: the substrate sequences for collagenase (MMP1, MMP3, MT-MMP, and others) and gelatinase (MMP2 and MMP9), as well as additional MTS sequences, were examined (MMP-G1/MTS, MMP-G2/MTS, MMP-G3/MTS, MMP-G4/MTS) (Fig. 6.3). Each sequence was introduced into the Fct14 expression plasmid, and its syncytium-inducing activity was examined in LLCMK2 cells by co-transduction with the HN-expressing plasmid and those expressing MMP2, MT1-MMP, or MT1-MMP/MMP2. The sequence

APRA↓MTS (↓: cleavage site, MMP-G4/MTS in Fig. 6.3) was found to give rise to the highest level of cell–cell fusion. A similar optimization approach was made for the uPA-targeted Fct14 sequence, revealing that the new motif, AGVSGR↓SFFGAV (insertion of S between R and F; F(uPA2), Fig. 6.3), was the most potent in terms of fusion induction. The recombinant Δ*M*SeV/Fct14s harboring those modified F proteins were generated and evaluated *in vivo* as described in the next section.

6.4 Studies Using Animal Models

6.4.1 In Vivo Proof-of-Concept Studies

For all the *in vivo* studies in this section, 7- to 8-week-old female balb/c nu/nu mice were used as the tumor cell-grafted models. Various human tumor cell lines (1×10^7 cells each) were subcutaneously injected into the right flank of the mice. When the subcutaneous tumors grew to 4–8 mm in diameter (usually 7 days after grafting), the mice were divided into groups ($n = 5$–8 per group), and each recombinant Δ*M*SeV/Fct14 [1×10^7 cell infectious units (CIU) in 100 μL phosphate-buffered saline (PBS)] was intratumorally injected. Tumor volumes were measured according to the formula $volume = a^2b/2$, where a is the shortest diameter and b is the longest diameter, and are expressed as the mean volume ± SE (Kinoh et al. 2009).

Δ*M*SeV/Fct14 (MMP-G4/MTS) was injected into the tumor tissues developed from human HT1080 and SW620 cells, respectively. As expected from the *in vitro* studies, the virus exhibited remarkable inhibition of the growth of MMP-expressing HT1080 tumor, but not that of SW620 expressing no MMPs (Fig. 6.4a). On the other hand (Fig. 6.4b), Δ*M*SeV/Fct14(uPA2)-GFP inhibited the growth of all three dermal tumors, CAKI-I, BxPC3, and PC14 (human cell lines from kidney clear cell carcinoma, pancreatic cancer, and lung adenocarcinoma, respectively), all expressing uPA at a high level. These inhibitions did not take place with MMP-targeted Δ*M*SeV/Fct14 (MMP-G4/MTS). No inhibition occurred, either, for HT29 (human colon adenocarcinoma cell line) tumor by either of the oncolytic viruses, likely because it expressed little or no detectable MMP and uPA (Kinoh et al. 2009). Thus, protease-specific or -dependent oncolytic virotherapy using those modified Δ*M*SeV/Fct14s appeared to be well validated *in vivo*.

We further examined the oncolytic (cytolytic) activity of those optimized Δ*M*SeVs using 33 established human tumor cell lines of various origins. Extensive cytolytic activity (LDH release of more than 40 %) of the uPA-type Δ*M*SeV [Δ*M*SeV/Fc14(uPA2)] was seen for 14 of the 33 tumor cell lines (42.4 %), although only 2 tumors (6.1 %) were susceptible to the MMP-targeted Δ*M*SeV [Δ*M*SeV/Fct14(MMP-G4/MTS)]. Taken together with a number of publications showing that the uPA system appears to be active in a majority of various tumors in situ (Egeblad and Werb 2002; Donà et al. 2004), our results suggest that the uPA-targeted Δ*M*SeV/Fct14 would be much more useful than MMP-targeted Δ*M*SeV/Fct14 for oncolytic virotherapy. Therefore, our subsequent studies preferentially used

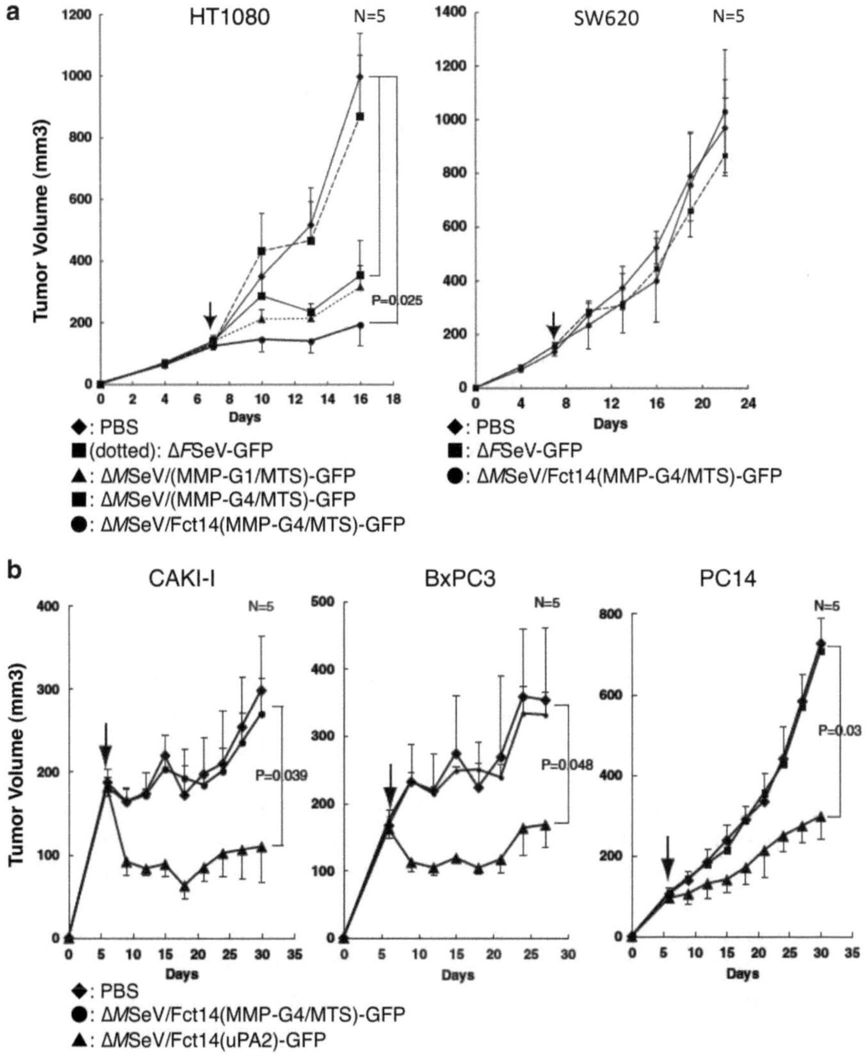

Fig. 6.4 Antitumor activity of optimized MMP- and uPA-targeted oncolytic Δ*M*SeV *in vivo*. (**a**) MMP-targeted Δ*M*SeV showed strong inhibition of grafted HT1080 tumor growth but no effect on the growth of SW620 tumors. (**b**) uPA-targeted Δ*M*SeV but not MMT-targeted Δ*M*SeV showed selective inhibition of *in vivo* growth of tumors, CAKI-I, BxPC3, and PC14 highly expressing uPA. For details, see text. (Adapted from Kinoh et al. 2009)

Δ*M*SeV/Fc14 (uPA2). This SeV was named "BioKnife" (Morodomi et al. 2012; reviewed in Morodomi et al. 2013), because it was expected to infect uPA-expressing tumors selectively and to eradicate them to the margin without damaging surrounding normal tissues. In other words, "BioKnife" appeared to represent a biological surgical weapon that could wipe out every tumor cell.

6.4.2 Case Studies Using BioKnife in Orthotopic Animal Models

6.4.2.1 Eradication of Glioblastomas with and without Synergism with Coexpressed Interferon-β in an Allograft Model

Background

Glioblastoma multiforme (GM), the most common primary malignant brain tumor, is highly invasive and intractable. The median survival of patients is merely 1 year with current standard treatments, including surgery and radiotherapy (DeAngelis 2001). Complete surgical removal of tumor cells without injuring surrounding brain tissues is extremely difficult. Moreover, the surgical margin is often regarded as positive because of the highly invasive nature of this tumor. A recent clinical trial showed that temozolomide, an oral alkylating agent, achieved significant improvement of the median survival of patients with GM to 14.6 months (Stupp et al. 2005), whereas another study reported that 73.5 % of patients still died within less than 2 years and 72.2 % experienced recurrence (Brandes et al. 2009). There is thus an urgent need for new approaches to GM treatment.

Virus-based therapeutics for glioblastoma have been evaluated in clinical settings during the past decade, including packaging cell-mediated local production of retroviruses expressing herpes simplex virus thymidine kinase (HSV-tk) (Rainov 2000). Although these viruses were shown to exert oncolytic activity and tumor specificity, no significant therapeutic effects were reported in the clinical trials (Liu et al. 2007a,b). Other approaches currently under investigation are using adenovirus, herpes simplex virus, and measles virus that have been manipulated to target GM specifically (Yokoyama et al. 2008; Wakimoto et al. 2009; Liu et al. 2007a, b; Piao et al. 2009).

It has been reported that GM frequently expresses uPA and that its invasiveness is closely related to this uPA activity (Rao 2003). Active uPA is generated from inactive pro-uPA on a high-affinity receptor, uPAR, on the cell surface. Then, uPA converts inactive plasminogen to active plasmin, resulting in the degradation of the ECM. Moreover, activation of these uPA-related proteins induces several intracellular signaling pathways via growth factor receptors, thereby facilitating cell adhesion, migration, and proliferation. A number of reports have demonstrated the close relationship between increased levels of local uPA and poor prognosis in tumor patients, including those bearing GM (Zhang et al. 2000). Because pro-uPA synthesis is preferentially upregulated in cancerous tissues but not in normal tissues and because active uPA is localized on the cancer cell surface, it seems reasonable to assume that a therapeutic modality targeting uPA could selectively kill cancer cells without damaging significantly the surrounding normal tissues.

These background features prompted us to investigate the potential of BioKnife, Δ*M*SeV/Fc14 (uPA2), for GM therapy.

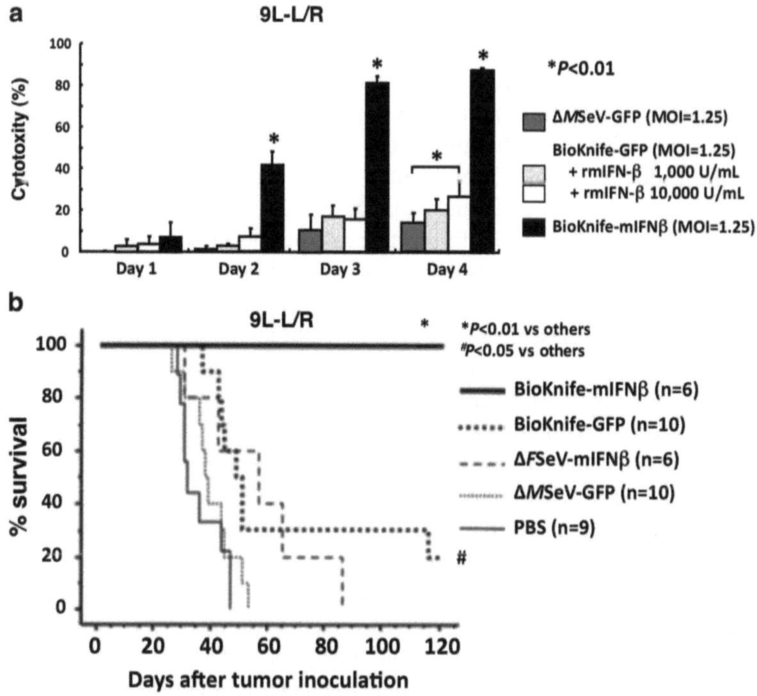

Fig. 6.5 Efficacy of BioKnife expressing mIFN-β gene *in vitro* and in the rat model for glioblastoma multiforme. (**a**) A synergistic effect was induced by BioKnife and its passenger gene, mIFN-β, in terms of killing of glioma 9L-L/R cells *in vitro*. (**b**) BioKnife-mIFN-β showed significant efficacy on an orthotopic rat brain tumor model grafted with 9L-L/R in terms of animal survival time. (Adapted from Hasegawa et al. 2010)

Experimental Results

We first observed uPA-specific cell fusion and killing by BioKnife for a number of human GM cell lines in culture. However, a rat gliosarcoma cell line, 9L, which we intended to use in a rat allograft model, was resistant to BioKnife-mediated cell death because the expression levels of uPA and uPAR were too low to be detectable in this cell line. We thus transformed 9L cells so that they would stably express the ligand uPA and the receptor uPAR. The transformed cells are here referred to as 9L-L/R. In addition, hoping for a synergistic effect, we inserted the gene encoding murine interferon-β (mIFN-β) into the *le/N* site (see Chap. 3) in BioKnife. The resulting product was named BioKnife-mIFN-β. BioKnife-mIFN-β was found to remarkably increase the cytotoxicity of 9L-L/R cells (Hasegawa et al. 2010) (Fig. 6.5a). Note that the addition of soluble mIFN-β did not enhance the cytotoxicity at all.

A total of 5×10^5 9L-L/R cells in 10 μL PBS were injected into the brain of male Fisher 344 rats using a microinjector over 5 min (for experimental details, see Iwadate et al. 2005). Briefly, rats were anesthetized with 50 mg/kg pentobarbital and placed in a stereotactic apparatus. A burr hole was made at an appropriate location (1 mm posterior of the bregma and 3 mm right of the midline). A 25-gauge needle was inserted at a point 3 mm ventral from the dura. On days 1, 4, and 7 after tumor injection, each BioKnife and other SeVs (2×10^7 CIU) in 20 μL PBS were injected in the same way. In addition, to evaluate the invasive activity, the same number of 9L-L/R cells in 10 μL PBS were injected into the brain of another rat in the same way and examined every 7 days by magnetic resonance imaging (MRI). The invasive activity was measured at 4 weeks after tumor inoculation. The rats were observed daily until severe paresis, ataxia, or periophthalmic encrustations developed or more than 20 % weight loss occurred.

As shown in Fig. 6.5b, all the control, tumor-bearing rats that received PBS or the parental Δ*M*SeV died within 46 and 52 days, respectively, whereas gene transfer of mIFN-β via Δ*F*SeV (for the structure of Δ*F*SeV, see Chap. 3) significantly improved the survival, likely because of an immune-mediated antitumor effect of IFN-β (Hasegawa et al. 2010). BioKnife alone also significantly improved survival. However, the therapeutic effect of BioKnife-mIFN-β was far more remarkable, rescuing all the tumor-bearing rats throughout the observation period. Thus, the results provided strong evidence for a profound synergistic effect between the oncolytic BioKnife and trans-expression of IFN-β on the therapy of GM in this rat orthotopic model.

6.4.2.2 Remarkable Elongation by BioKnife of the Survival Time of Animals with Pleurally Spreading Malignant Mesothelioma

Background

Several decades ago, asbestos was widely used for its industrial and economic advantages, non-conductiveness, thermal insulation property, and sound absorbability. Epidemiological and pathological studies, however, subsequently revealed its tumorigenic potential (Stanton et al. 1981; Marinaccio et al. 2007; Murayama et al. 2006). As a result, the use of asbestos was prohibited in almost all developed countries. Because asbestos-induced tumors occur several decades after exposure, the rapidly increasing number of asbestos-related malignancies has now become a serious issue. Malignant pleural mesothelioma (MPM) is one of those asbestos-related malignancies arising from the pleural cavity, and is highly intractable and resistant to the current standard therapeutics. In fact, MPM has a median overall survival rate of less than 30 months, even when multimodality therapy is used (Sugarbaker et al. 1999; Bölükbas et al. 2011). Therefore, novel therapeutics for this condition is urgently needed.

Experimental Results

We have assessed the therapeutic potential of BioKnife for MPM using orthotopic xenograft models. The *in vivo* MPM orthotopic xenograft models were established according to Martarelli et al. (2006). We used two types of human MPM cell line, H226 (epithelioid subtype) and MSTO-211H (biphasic subtype), and collected cells in the logarithmic growth phase (Morodomi et al. 2012). They were grafted into the right thoracic cavity of 4-week-old male balb/c *nu/nu* mice at two different doses (1×10^6 and 5×10^6 cells in 100 μL PBS/head). Both H226 and MSTO-211H cells developed into some tumor nodules, which later increased in number in both of the pleural cavities (Morodomi et al. 2012). For either cell type, 5×10^6 cells were considered to be an appropriate dose, because all the mice administered this dose were killed within 128 days by H226 and within 38 days by MSTO-211H. When tumor cells were accidentally injected into the subcutaneous space or when mice showed pneumothorax or hemothorax, the animals were killed by excess anesthesia and excluded from further analysis. The day of cell inoculation into mice was regarded as day 0. Seven days later, when the tumor nodules were established, an appropriate amount of BioKnife or PBS was administered into the left pleural cavity.

Next, to assess the therapeutic effect of BioKnife, we examined the dose–response relationship in the tumor-bearing mice via single intrapleural administration of BioKnife on day 7 at 4×10^5, 2×10^6, and 1×10^7 CIU/dose, when the tumors were established. Significant survival prolongation was observed in both tumor models when a dose of 1×10^7 CIU was used. Therefore, this dose was used in the subsequent experiments (Morodomi et al. 2012).

We performed single and multiple administrations of BioKnife after tumor cell inoculation, once on day 7, three times on days 7, 9, and 11, and six times on days 7, 9, 11, 13, 15, and 17. The highest and most noteworthy therapeutic effect was achieved by the use of 6 injections of BioKnife in both H226 and MSTO-211H models (Morodomi et al. 2012) (Fig. 6.6).

The significance of these results was further substantiated by the following observations. First, all the nine surgical specimens from patients with mesothelioma expressed uPAR. Second, although H229 cells were found to be highly susceptible to BioKnife, they expressed little uPA. It was suggested that BioKnife can stimulate uPA expression via the RNA helicase RIG-I and the NF-κB-dependent pathway. It thus appeared that uPA expression itself is not always particularly important, whereas uPAR expression could be a prerequisite for BioKnife-mediated cancer cell killing.

6.5 Conclusions

Oncolytic virotherapy is not a very new but rather a reemerging approach to cancer therapy. Diverse virus species are now under investigation for the development of a more cancer-specific powerful tool. Our studies have provided the proof-of-concept

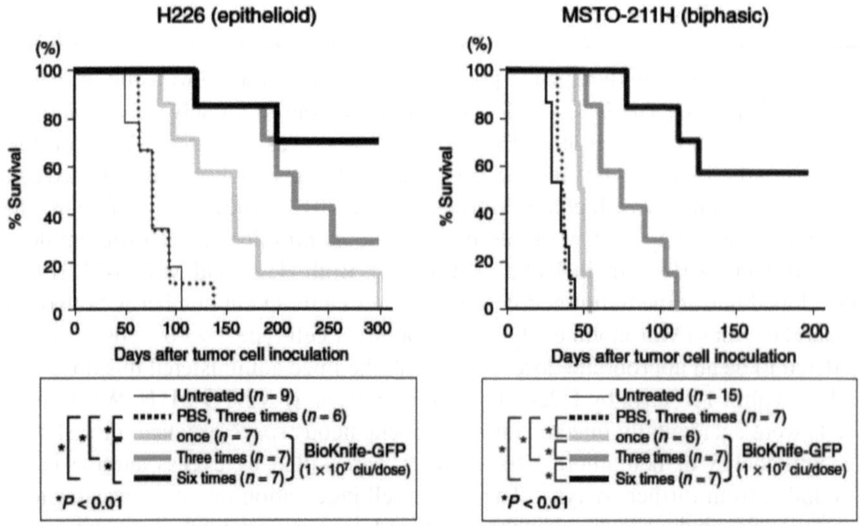

Fig. 6.6 Efficacy of BioKnife in the malignant pleural mesothelioma grafted model. Multiple administrations of BioKnife showed significant efficacy (extension of survival time) on the orthotopic models of two independent malignant pleural mesotheliomas (H226, epithelioid subtype; and MSTO-211H, biphasic subtype) *in vivo*. (Adapted from Morodomi et al. 2012)

for the feasibility of using Δ*M*SeV with profound cell–cell fusion and cell-killing capacity, which is further engineered to selectively target malignant tumor cells by applying the protease-dependent theory of SeV tissue tropism, thus proposing a novel concept for oncolytic virotherapy. The uPA-targeted BioKnife appeared to be especially useful for its sharpness and wide applicability. Further studies are expected to determine whether this novel modality is safe and effective in clinical settings.

Acknowledgments We are grateful to M. Inoue at DNAVEC Corporation for his critical reading of the manuscript.

References

Andreasen PA, Egelund R, Petersen HH (2000) The plasminogen activation system in tumor growth, invasion, and metastasis. Cell Mol Life Sci 57:25–40

Bölükbas S, Manegold C, Eberlein M, Bergmann T, Fisseler-Eckhoff A, Schirren J (2011) Survival after trimodality therapy for malignant pleural mesothelioma: radical pleurectomy, chemotherapy with cisplatin/pemetrexed and radiotherapy. Lung Cancer 71:75–81. doi:10.1016/j.lungcan.2009.08.019

Brandes AA, Tosoni A, Franceschi E, Sotti G, Frezza G, Amistà P, Morandi L, Spagnolli F, Ermani M (2009) Recurrence pattern after temozolomide concomitant with and adjuvant to radiotherapy in newly diagnosed patients with glioblastoma: correlation with MGMT promoter methylation status. J Clin Oncol 27:1275–1279. doi:10.1200/JCO.2008.19.4969

Cathomen T, Naim HY, Cattaneo R (1998) Measles viruses with altered envelope protein cytoplasmic tails gain cell fusion competence. J Virol 72:1224–1234

Chiocca EA (2002) Oncolytic viruses. Nat Rev Cancer 2:938–950

Cox G, O'Byrne KJ (2001) Matrix metalloproteinases and cancer. Anticancer Res 21:4207–4219

Curran S, Murray GI (1999) Matrix metalloproteinases in tumour invasion and metastasis. J Pathol 189:300–308

DeAngelis LM (2001) Brain tumors. N Engl J Med 344:114–123

Donà M, Dell'Aica I, Pezzato E, Sartor L, Calabrese F, Della Barbera M, Donella-Deana A, Appendino G, Borsarini A, Caniato R, Garbisa S (2004) Hyperforin inhibits cancer invasion and metastasis. Cancer Res 64:6225–6232

Egeblad M, Werb Z (2002) New functions for the matrix metalloproteinases in cancer progression. Nat Rev Cancer 2:161–174

Friedberg MH, Glantz MJ, Klempner MS, Cole BF, Perides G (1998) Specific matrix metalloproteinase profiles in the cerebrospinal fluid correlated with the presence of malignant astrocytomas, brain metastases, and carcinomatous meningitis. Cancer (Phila) 82:923–930

Fujiwara T, Urata Y, Tanaka N (2008) Diagnostic and therapeutic application of telomerase-specific oncolytic adenoviral agents. Front Biosci 13:1881–1886

Gotoh B, Ogasawara T, Toyoda T, Inocencio NM, Hamaguchi M, Nagai Y (1990) An endoprotease homologous to the blood clotting factor X as a determinant of viral tropism in chick embryo. EMBO J 9:4189–4195

Hasegawa Y, Kinoh H, Iwadate Y, Onimaru M, Ueda Y, Harada Y, Saito S, Furuya A, Saegusa T, Morodomi Y, Hasegawa M, Saito S, Aoki I, Saeki N, Yonemitsu Y (2010) Urokinase-targeted fusion by oncolytic Sendai virus eradicates orthotopic glioblastomas by pronounced synergy with interferon-β gene. Mol Ther 18:1778–1786. doi:10.1038/mt.2010.138

Homma M, Ouchi M (1973) Trypsin action on the growth of Sendai virus in tissue culture cells. 3. Structural difference of Sendai viruses grown in eggs and tissue culture cells. J Virol 12:1457–1465

Inoue M, Tokusumi Y, Ban H, Kanaya T, Shirakura M, Tokusumi T, Hirata T, Nagai Y, Iida A, Hasegawa M (2003) A new Sendai virus vector deficient in the matrix gene does not form virus particles and shows extensive cell-to-cell spreading. J Virol 77:6419–6429

Iwadate Y, Inoue M, Saegusa T, Tokusumi Y, Kinoh H, Hasegawa M, Tagawa M, Yamaura A, Shimada H (2005) Recombinant Sendai virus vector induces complete remission of established brain tumors through efficient interleukin-2 gene transfer in vaccinated rats. Clin Cancer Res 11:3821–3827

Kelly E, Russell SJ (2007) History of oncolytic viruses genesis to genetic engineering. Mol Ther 15:651–659

Kinoh H, Inoue M, Washizawa K, Yamamoto T, Fujikawa S, Tokusumi Y, Iida A, Nagai Y, Hasegawa M (2004) Generation of a recombinant Sendai virus that is selectively activated and lyses human tumor cells expressing matrix metalloproteinases. Gene Ther 11:1137–1145

Kinoh H, Inoue M, Komaru A, Ueda Y, Hasegawa M, Yonemitsu Y (2009) Generation of optimized and urokinase-targeted oncolytic Sendai virus vectors applicable for various human malignancies. Gene Ther 16:392–403. doi:10.1038/gt.2008.167

Liu C, Sarkaria JN, Petell CA, Paraskevakou G, Zollman PJ, Schroeder M, Carlson B, Decker PA, Wu W, James CD, Russell SJ, Galanis E (2007a) Combination of measles virus virotherapy and radiation therapy has synergistic activity in the treatment of glioblastoma multiforme. Clin Cancer Res 13:7155–7165

Liu TC, Galanis E, Kirn D (2007b) Clinical trial results with oncolytic virotherapy: a century of promise, a decade of progress. Nat Clin Pract Oncol 4:101–117

Marinaccio A, Binazzi A, Cauzillo G, Cavone D, Zotti RD, Ferrante P, Gennaro V, Gorini G, Menegozzo M, Mensi C, Merler E, Mirabelli D, Montanaro F, Musti M, Pannelli F, Romanelli A, Scarselli A, Tumino R, Italian Mesothelioma Register (ReNaM) Working Group (2007) Analysis of latency time and its determinants in asbestos related malignant mesothelioma cases of the Italian Register. Eur J Cancer 43:2722–2728

Martarelli D, Catalano A, Procopio A, Orecchia S, Libener R, Santoni G (2006) Characterization of human malignant mesothelioma cell lines orthotopically implanted in the pleural cavity of immunodeficient mice for their ability to grow and form metastasis. BMC Cancer 6:130

Morodomi Y, Yano T, Kinoh H, Harada Y, Saito S, Kyuragi R, Yoshida K, Onimaru M, Shoji F, Yoshida T, Ito K, Shikada Y, Maruyama R, Hasegawa M, Maehara Y, Yonemitsu Y (2012) BioKnife, a uPA activity-dependent oncolytic Sendai virus, eliminates pleural spread of malignant mesothelioma via simultaneous stimulation of uPA expression. Mol Ther 20:769–777. doi:10.1038/mt.2011.305

Morodomi Y, Inoue M, Hasegawa M, Okamoto T, Maehara Y, Yonemitsu Y (2013) Sendai virus-based oncolytic gene therapy. In: Novel gene therapy approaches. Ming Wei and David Good (Eds.), chapter 9, pp 183–194. InTech, HYPERLINK "http://doi.org/10.5772/55328" \t "_blank" doi.org/10.5772/55328

Murayama T, Takahashi K, Natori Y, Kurumatani N (2006) Estimation of future mortality from pleural malignant mesothelioma in Japan based on an age-cohort model. Am J Ind Med 49:1–7

Nagai Y (1993) Protease-dependent virus tropism and pathogenicity. Trends Microbiol 1:81–87, Erratum in Trends Microbiol (1993) 1:135

Ogasawara T, Gotoh B, Suzuki H, Asaka J, Shimokata K, Rott R, Nagai Y (1992) Expression of factor X and its significance for the determination of paramyxovirus tropism in the chick embryo. EMBO J 11:467–472

Piao Y, Jiang H, Alemany R, Krasnykh V, Marini FC, Xu J, Alonso MM, Conrad CA, Aldape KD, Gomez-Manzano C, Fueyo J (2009) Oncolytic adenovirus retargeted to Delta-EGFR induces selective antiglioma activity. Cancer Gene Ther 16:256–265. doi:10.1038/cgt.2008.75

Rainov NG (2000) A phase III clinical evaluation of herpes simplex virus type 1 thymidine kinase and ganciclovir gene therapy as an adjuvant to surgical resection and radiation in adults with previously untreated glioblastoma multiforme. Hum Gene Ther 11:2389–2401

Rao JS (2003) Molecular mechanisms of glioma invasiveness: the role of proteases. Nat Rev Cancer 3:489–501

Rein A, Mirro J, Haynes JG, Ernst SM, Nagashima K (1994) Function of the cytoplasmic domain of a retroviral transmembrane protein: p15E-p2E cleavage activates the membrane fusion capability of the murine leukemia virus Env protein. J Virol 68:1773–1781

Richardson CD, Scheid A, Choppin PW (1980) Specific inhibition of paramyxovirus and myxovirus replication by oligopeptides with amino acid sequences similar to those at the N-termini of the F1 or HA2 viral polypeptides. Virology 105:205–222

Ring CJ (2002) Cytolytic viruses as potential anti-cancer agents. J Gen Virol 83:491–502

Russell SJ, Peng KW, Bell JC (2012) Oncolytic virotherapy. Nat Biotechnol 30:658–670

Stanton MF, Layard M, Tegeris A, Miller E, May M, Morgan E, Smith A (1981) Relation of particle dimension to carcinogenicity in amphibole asbestoses and other fibrous minerals. J Natl Cancer Inst 67:965–975

Stupp R, Mason WP, van den Bent MJ, Weller M, Fisher B, Taphoorn MJ, Belanger K, Brandes AA, Marosi C, Bogdahn U, Curschmann J, Janzer RC, Ludwin SK, Gorlia T, Allgeier A, Lacombe D, Cairncross JG, Eisenhauer E, Mirimanoff RO, European Organisation for Research and Treatment of Cancer Brain Tumor and Radiotherapy Groups, National Cancer Institute of Canada Clinical Trials Group (2005) Radiotherapy plus concomitant and adjuvant temozolomide for glioblastoma. N Engl J Med 352:987–996

Sugarbaker DJ, Flores RM, Jaklitsch MT, Richards WG, Strauss GM, Corson JM, DeCamp MM Jr, Swanson SJ, Bueno R, Lukanich JM, Baldini EH, Mentzer SJ (1999) Resection margins, extrapleural nodal status, and cell type determine postoperative long-term survival in trimodality therapy of malignant pleural mesothelioma: results in 183 patients. J Thorac Cardiovasc Surg 117:54–63, discussion 63–65

Todo T (2012) Active immunotherapy: oncolytic virus therapy using HSV-1. Adv Exp Med Biol 746:178–186

Wakimoto H, Kesari S, Farrell CJ, Curry WT Jr, Zaupa C, Aghi M, Kuroda T, Stemmer-Rachamimov A, Shah K, Liu TC, Jeyaretna DS, Debasitis J, Pruszak J, Martuza RL, Rabkin SD (2009) Human glioblastoma-derived cancer stem cells: establishment of invasive glioma models and treatment with oncolytic herpes simplex virus vectors. Cancer Res 69:3472–3481. doi:10.1158/0008-5472.CAN-08-3886

Yokoyama T, Iwado E, Kondo Y, Aoki H, Hayashi Y, Georgescu MM, Sawaya R, Hess KR, Mills GB, Kawamura H, Hashimoto Y, Urata Y, Fujiwara T, Kondo S (2008) Autophagy-inducing agents augment the antitumor effect of telerase-selve oncolytic adenovirus OBP-405 on glioblastoma cells. Gene Ther 15:1233–1239. doi:10.1038/gt.2008.98

Zhang X, Fei Z, Bu X, Zhen H, Zhang Z, Gu J, Chen Y (2000) Expression and significance of urokinase type plasminogen activator gene in human brain gliomas. J Surg Oncol 74:90–94

Chapter 7
Induction of Human Pluripotent Stem Cells by the Sendai Virus Vector: Establishment of a Highly Efficient and Footprint-Free System

Noemi Fusaki and Hiroshi Ban

Abstract The Sendai virus (SeV) vector system exhibited very high performance in transgene expression and a broad target tissue/cell range. Together with its non-integrating nature, the potential of this vector has been suggested to be an efficient and valuable tool for generating induced pluripotent stem cells (iPSCs). Initial stage studies have shown the potential of the SeV vector and demonstrated that this system is simple and generates vector/transgene-free iPSCs with high efficiency. Subsequently, a more controllable method to eliminate the vector/transgenes from generated iPSCs has been developed by introducing several temperature-sensitive mutations into the SeV vector backbone. Based on this method, human iPSCs have been established from cord blood CD34$^+$ cells, T cells from peripheral blood, and samples of patients with intractable diseases. In addition, a model of regenerative medicine with gene correction was presented using SeV vector-derived iPSCs from a patient with an inherited disease.

7.1 Introduction

The discovery that human somatic cells are reprogrammed to generate induced pluripotent stem cells (iPSCs) by the intracellular delivery and expression of a combination of four special reprogramming factors, namely, *OCT3/4*, *SOX2*, *KLF4*, and

N. Fusaki (✉)
Department of Ophthalmology, Keio University School of Medicine, 35 Shinanomachi, Shinjuku-ku, Tokyo 160-8582, Japan

Precursory Research for Embryonic Science and Technology, Japan Science and Technology Agency, 4-1-8 Honcho Kawaguchi, Saitama 332-0012, Japan

DNAVEC Corporation, 6 Ohkubo, Tsukuba-shi, Ibaraki 300-2611, Japan
e-mail: nfusaki@a2.keio.jp

H. Ban
DNAVEC Corporation, 6 Ohkubo, Tsukuba-shi, Ibaraki 300-2611, Japan

Y. Nagai (ed.), *Sendai Virus Vector: Advantages and Applications*,
DOI 10.1007/978-4-431-54556-9_7, © Springer Japan 2013

c-*MYC*, has opened the gate to a new and exciting era in the vast field of life sciences, that is, cell biology, developmental biology, and future expected regenerative medicine (Takahashi and Yamanaka 2006; Yamanaka 2010). However, the introduction of these reprogramming factors into cells has mostly been conducted using retrovirus (including lentivirus) vectors that are integrated into host chromosomes, which may be one of the factors increasing the risk of tumorigenesis (Okita et al. 2007). In addition, use of the c-*MYC* gene, one of the four reprogramming factors, has raised concern because it is known to be potentially oncogenic. Thus, safer reprogramming methods with sufficiently high efficiency have been extensively explored, including the use of plasmids (Okita et al. 2008; Yu et al. 2009), the Cre/loxP system (Soldner et al. 2009; Sommer et al. 2009), adenovirus vectors (Stadtfeld et al. 2008), the PiggyBac technology (Woltjen et al. 2009; Kaji et al. 2009), the minicircle vector (Jia et al. 2010), the proteins (Zhou et al. 2009; Kim et al. 2009), and the synthetic modified mRNAs (Warren et al. 2010). However, low efficiency has been associated with these methods, which sometimes require the repetitive introduction of reprogramming factors. Insufficient excision of vectors is also a problem. Synthetic modified mRNA may represent a solution to these problems; however, the reagents must be added to the target cells every day (Warren et al. 2010) and its reprogramming efficiency is still not so high. Thus, new, highly efficient, and simpler reprogramming methods that fulfill the foregoing crucial requirements have been a pressing need for not only basic research that should not be affected by chromosomal integration or remaining vectors, but also for the safer clinical application of iPSCs in the near future.

One of the unique characteristics intrinsic to Sendai virus (SeV) is that the virus lifecycle is completed entirely in the cytoplasm as an RNA form, with no DNA phase of nuclear localization; therefore, there is theoretically no risk of chromosomal integration (see Chap. 2). The expression levels of foreign genes from the replicating SeV vector backbone in the cytoplasm are very high, and more importantly, are controllable by changing the insertion site of transgenes in the vector (see Chap. 3). Furthermore, various versions of nontransmissible SeV vectors have been created by deleting one or more genes coding for the viral structural proteins, which are safer than the fully transmissible original SeV vector with the full-length viral genome (see Chap. 3). These advances in engineering along with the nonintegrating nature would make SeV vector a useful and powerful tool for the generation of iPSCs without being associated with the concerns just described.

In this chapter, we describe the establishment of highly efficient and simple protocols utilizing newly designed SeV vectors to generate footprint-free iPSCs, that is, iPSCs devoid of the vector and transgenes that have been employed for reprogramming.

7.2 Generation of Footprint-Free iPSCs Using the SeV Vector: The Initial Studies

To learn whether the SeV vector would be able to generate iPSCs from human skin fibroblasts, a vector derived from the nontransmissible *F* gene-deleted vector backbone Δ*F*SeV (Fig. 7.1a; Li et al. 2000; also see Chap. 3) was employed. This vector

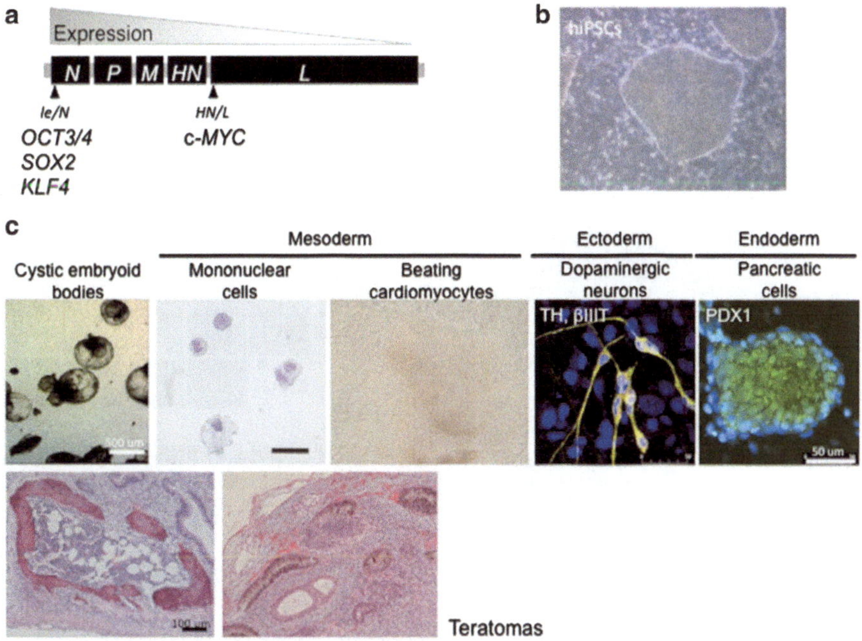

Fig. 7.1 (**a**) Schematic representation of the Δ*F*SeV TS conv vector. The expression level of an inserted gene decreased depending on the insertion site toward the 5′-end of the viral genome (polar attenuation). Reprogramming genes were inserted into *le/N* site (*OCT3/4*, *SOX2*, and *KLF4*) or *HN/L* site (c-*MYC*), giving rise to the highest efficiency of induced pluripotent stem cell (iPSC) generation (see text). (**b**) Morphology of the obtained iPSCs. (**c**) Pluripotency of the established iPSCs. The generated iPSCs formed cystic embryoid bodies and differentiated into the mesoderm (mononuclear cells and beating cardiomyocytes), ectoderm (dopaminergic neurons) that expressed tyrosine hydroxylase (*TH*) and βIII tubulin (*βIIIT*) represented by the *yellow lines*, and endoderm (pancreatic cells positive for *PDX1* in *green*). The iPSCs also differentiated into various tissues including bone marrow-like tissue (*left*), muscles, and neuron-like structures (*right*) in teratomas formed in NOD/SCID mice. (Adapted from Fusaki et al. 2009)

was created to achieve further reduction in the *in vitro* cytopathogenicity of the Δ*F*SeV vector by introducing the following temperature-sensitive (ts) point mutations: G69F, T116A, and A183S in the M protein, which were derived from the persistently infectious SeV clone 151, and A262T, G264R, and K461G in the HN protein (Inoue et al. 2003; see Chaps. 2 and 3), in addition to L511F in the P protein, and N1197S and K1795E in the L proteins (Yoshizaki et al. 2006). This new construct was named "TS conv," the conventional ts vector, and used in the initial stage of this reprogramming study (see Fig. 7.2a). Transduction, as measured using the green fluorescent protein (GFP) encoded on the TS conv, showed that almost all infected human dermal fibroblast cells expressed GFP at a low multiplicity of infection (MOI) [three cell infectious units (CIU) per cell] (Fusaki et al. 2009).

We employed the strategy to insert each of the four reprogramming factors, *OCT3/4*, *SOX2*, *KLF4*, and c-*MYC* (Takahashi et al. 2007), into a separate, single

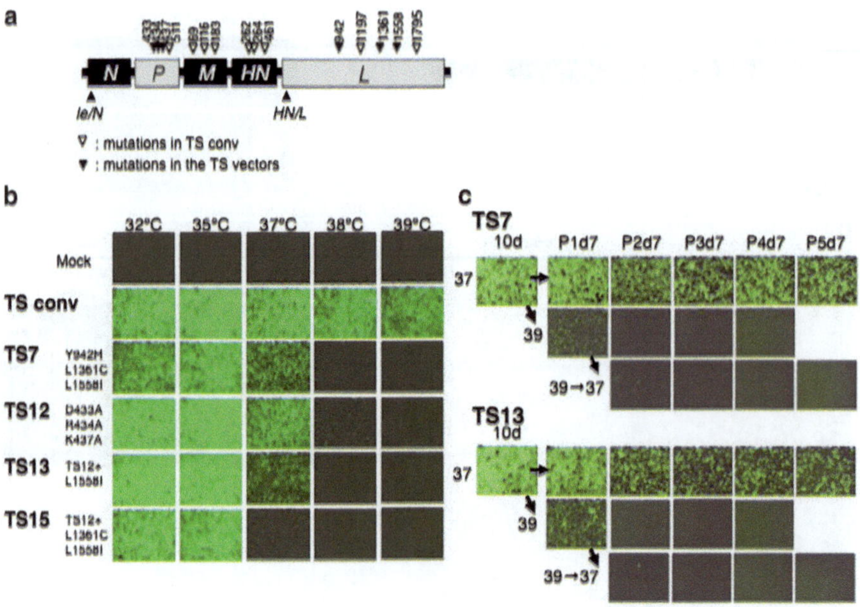

Fig. 7.2 (**a**) Positions of temperature-sensitive (ts) mutations in the TS conv (*open triangles*) and additional ts mutations introduced to generate new TS vectors (*closed triangles*). (**b**) Temperature sensitivity of reporter (GFP) gene expression from TS conv and new TS vectors with the additional ts mutations at different temperatures. For details, see text. Green fluorescence was compared 3 days after infection. (**c**) The effect of temperature shift-up and shift-down on GFP expression from TS7 and TS 13. For details, see text. (Adapted from Ban et al. 2011)

TS conv vector rather than to insert the four factors altogether into a single vector. The reasons for this are as follows.

First, the maximum length of a foreign gene that can be accommodated by Δ*F*SeV is 5.4 kb (Ban et al. 2007; see Chap. 3). In addition, when the insertion is longer, the vector production titers become significantly lower (Sakai et al. 1999). The length of the four reprogramming factors in tandem with the transcription unit for each of them accounted for 5.0 kb, nearly the upper limit of accommodation by Δ*F*SeV. These situations raised the fear that insertion of the four factors into a single vector would result in a greatly reduced production titer of the vector stock. Second, the vector production cells would be converted to iPS-like cells or undergo unforeseen change if such a single construct encoding the four factors is transfected. Third, optimization of reprogramming would require the determination of the optimum ratio of the four factors to be supplied. This study obviously cannot be done using a single vector system and is possible only under the conditions where the vectors expressing each of the four factors are given in various quantitative combinations. Finally, it would be necessary to protect researchers from directly contacting the four reprogramming factors at once, which were on the TS conv with very high transduction efficiency, a common nature of the SeV vectors.

As described in Chap. 3, the SeV vector houses several insertion sites for passenger genes (transgenes). The more downstream is the insertion site, the lower is the expression level (polar attenuation; Tokusumi et al. 2002). Thus, the level is the highest from the most upstream *le/N* site, whereas the lowest from the most downstream *L/tr* site (for the designation of insertion sites, see Chap. 3) (Fig. 7.1a); also see Chap. 3 for the details of insertion with the transcription unit. We attempted to optimize the insertion sites of the four reprogramming factors and determine the ratio of combination of these four recombinant TS conv vectors to achieve the most efficient generation of iPSCs. One of the best combinations was found to use the TS conv vectors with the insertion of *OCT3/4*, *SOX2*, and *KLF4* at the *le/N* site and c-*MYC* at the *HN/L* site (Fig. 7.1a). Detailed conditions for the generation of human iPSCs were previously described (Fusaki et al. 2009). Briefly, a neonatal human foreskin-derived fibroblast cell line (BJ) or adult human facial dermal fibroblast cell line (HDF) was plated onto plastic dishes and infected with the foregoing recombinant four TS conv vectors at an MOI of 3. One week after infection, cells were replated on mitomycin C-treated MEF feeder cells. The growth medium was replaced the next day with primate ES medium supplemented with human basic fibroblast growth factor. Reprogramming efficiency was defined by the ratio of the number of alkaline phosphatase (ALP)-positive colonies detected to the total number of infected cells seeded. The reprogramming efficiency was about 1 % for BJ (Fusaki et al. 2009). The morphologies of these colonies resembled that of human ES cells (hESCs) (Fig. 7.1a).

The TS conv vectors in the resultant cells were diluted out during subsequent passages with the robust division of cells, which led to the establishment of iPSC cell lines. In an alternative method, vector-positive cells were selectively eliminated using an anti-SeV-HN antibody that recognized the viral protein HN expressed on the surface of these cells (Fusaki et al. 2009). These transgene-free iPSCs expressed the hES markers, formed cystic embryoid bodies, exhibited pluripotency to differentiate into human tissues with three different germinal layers *in vitro*, and produced teratomas *in vivo* when transplanted in immunodeficient mice (Fig. 7.1c). It was confirmed that such iPSCs obtained by the SeV vectors did not contain any SeV vector genomes or any additional copies of *OCT3/4*, *SOX2*, *KLF4*, and c-*MYC* other than their intrinsic presence in the chromosomes. These results demonstrated that the SeV vector system worked well and very efficiently generated human iPSCs without chromosomal integration, and that these cells were completely free of vectors and reprogramming transgenes. In other words, this SeV vector method is an excellent footprint-free system for iPSC generation. It should be noted that the method is of value because of its simplicity; that is, only a single transduction of target somatic cells was enough to generate human iPSCs.

More recently, Nishimura et al. (2011) also succeeded in iPSCs generation from mouse primary fibroblasts using an SeV vector. In this study, a replication defective and poorly cytopathic SeV clone (cl. 151) isolated from persistently infected cells (see Chap. 2) was used as the vector backbone, and the four reprogramming factors were installed in a single vector. However, siRNA appeared to be necessary to erase the vector and the *trans*-genes.

7.3 Establishment of a More Controllable Footprint-Free System for iPSCs Generation

The method to obtain footprint-free iPSCs in the foregoing section depends on a spontaneous decrease in the vector in the resultant reprogrammed cells. Although the system works well for many types of human cells, some cell types require a longer time and many passages of the reprogrammed cells to achieve complete vector removal. This restriction led to trials to establish a more controllable method to remove the vectors as well as the reprogramming factors from the iPSCs generated.

7.3.1 Creation of New SeV Vectors by Introducing Additional Temperature-Sensitive Mutations

The shutoff of gene expression from the TS conv vector by shifting the culture temperature (37–38 °C or 39 °C) was found to be incomplete (see Fig. 7.2b). To establish a more clear-cut shutoff vector system, we attempted to introduce additional ts mutations into the TS conv vector backbone that had been identified in the viral polymerase complex, the L and P proteins of ts SeV (Bowman et al. 1999; Feller et al. 2000) (Fig. 7.2a). Because of the error-prone nature of RNA viruses, a single point mutation may allow them to revert to the wild type more readily than multiple point mutations. Therefore, we introduced at least three ts mutations in combination into the TS conv vector.

The new TS vector constructs generated by introducing the new ts mutations (shown in parentheses) were the TS7 vector (Y942H, L1361C, and L1558I in the L protein), TS12 vector (D433A, R434A, and K437A in the P protein), and TS13 vector (L1558I in the L protein), in addition to the same three mutations in the TS12 vector, and TS15 vector (L1361C and L1558I in the L protein) in addition to the same three mutations in the TS12 vector (Ban et al. 2011) (Fig. 7.2a, b).

All the TS vectors including the TS conv were derived from Δ*F*SeV. The reporter gene encoding GFP was positioned in place of the original *F* gene. The gene expression profiles of the reporter GFP gene from these new TS vectors were compared with that of the TS conv vector at various culture temperatures from 32 to 39 °C in LLC-MK2 cells (Fig. 7.2b). The shutoff of gene expression was clear cut at 38–39 °C for the TS7, TS12, and TS13 vectors, although gene expression was already significantly attenuated at 37 °C. The shutoff temperature was 37 °C for the TS15 vector (Fig. 7.2b). Cells transfected with the TS7 vector and TS13 vector were cultured at 37 °C for 10 days and then split into two groups, respectively. One of the groups was continually cultured at 37 °C and the other was cultured at an elevated temperature of 39 °C. Cells from both groups were subcultured every 7 days until the fourth or fifth passage (a total of 28 and 35 days, respectively). GFP expression

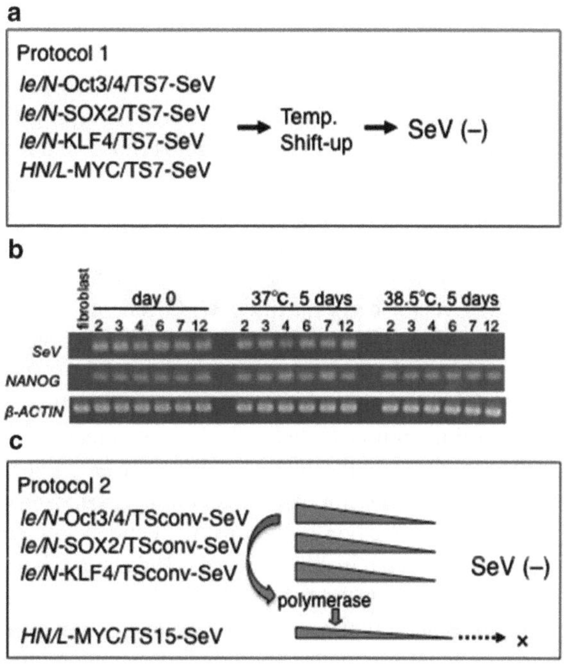

Fig. 7.3 The two different protocols, 1 (**a**) and 2 (**c**), employed to generate human iPSCs with different TS versions. For details, see text. (**b**) Detection of the SeV genome by RT-PCR in Protocol 1. For details, see text. (Adapted from Ban et al. 2011)

intensity was greatly reduced after one passage (P1d7) at 39 °C and became undetectable for both the TS7 and TS13 vectors (Fig. 7.2c). When the temperature was reversely shifted down from 39 to 37 °C and cells were cultured further up to 7 days, no GFP fluorescence reemerged. None of these TS vectors was cytotoxic, and all infected cells remained attached to the culture dishes at all the temperatures tested, indicating that functional vectors had disappeared during the culture at 39 °C, a nonpermissive temperature.

7.3.2 *Protocols for Generating Human iPSCs Using the Newly Developed TS Vectors*

By employing these newly developed TS vectors, we devised two different protocols to efficiently generate footprint-free human iPSCs. The insertion sites of the four reprogramming genes were identical to those used in the TS conv (Fig. 7.3a,c; cf. Figs. 7.1a, 7.2a). In the first protocol, the TS7 vector was used as the vector backbone (Fig. 7.3a, Protocol 1) (Ban et al. 2011). After establishing iPSC clones (day 0) at a permissive temperature of 37 °C, the clones were further cultured for 5 days at 37 °C or at a nonpermissive temperature of 38.5 °C. Although the SeV vector genome was detected in all the six clones cultured in the former conditions, none of the vectors was detected in the clones cultured in the latter conditions (Fig. 7.3b). Thus, as expected, the TS7 vector as well as the passenger transgenes (reprogramming factors) in iPSCs could be erased after reprogramming by controlling the culture temperature, shifting up to a nonpermissive temperature. However, because of the attenuation of gene expression of the TS7 vector at 37 °C, a tenfold-higher MOI (30 CIU per cell) was required in this protocol.

In the second protocol (Fig. 7.3c, Protocol 2), the three TS conv vectors carrying the *OCT3/4, SOX2* and *KLF4* genes, respectively, were used in combination with the TS15 vector carrying the c-*MYC* gene at a cell culture temperature of 37 °C without shifting (Ban et al. 2011). This idea was based upon the following observations and speculations.

The TS conv with the c-*MYC* gene inserted at its *HN/L* site always persisted for a longer time in transfected cells than any other TS conv vectors carrying *OCT3/4*, *SOX2*, or *KLF4* at the *le/N* site. When inserted at the *le/N* position, the c-M*YC*-carrying vector did not display such longer persistence. Thus, the life span of a TS conv with an insertion at the upstream site (*le/N*) appeared to be shorter than that with a downstream insertion into the *HN/L* site. Then, in coinfection with these two different TS convs, the former would be gradually outcompeted by the latter. TS15 was incapable of replication and gene expression even at 37 °C, while this temperature is well permissive for TS convs, allowing vigorous gene expression (Fig. 7.2b). Thus, under coinfection conditions, TS15 with the *cMYC* at *HN/L* site could be readily rescued to replicate and express genes at 37 °C by the RNA polymerase expressed by TS convs (Fig. 7.3c). It is thus likely that so long as the RNA polymerase is supplied by TS convs, TS15 is not only active as an expression vector but also likely capable of outcompeting the TS convs expressing the other three reprogramming factors. We expected that TS convs would disappear in due course of time, no longer supplying the RNA polymerase, and accordingly TS15 would also become extinct.

Indeed, this has taken place (Ban et al. 2011). In addition, the robust gene expression from the TS conv vectors at 37 °C (Fig. 7.2b) suggested that a lower MOI was sufficient to generate iPSCs in this second protocol than that (MOI 30; see above) in the first protocol using TS7 vectors whose replication and transgene expression at 37 °C were markedly attenuated (Fig. 7.2b); this was also true. Actually, MOI 3 was found to be sufficient in Protocol 2 to generate alkaline phosphatase (ALP)-positive colonies with human embryonic stem (ES) cell-like morphology from human fibroblast cells 28 days after transduction. Although the SeV vectors and transgenes were still positive in these cells, they became undetectable by reverse transcription-polymerase chain reaction (RT-PCR) and immunoassays after the subsequent passages over 2 or 3 months. It is practically of additional merit that Protocol 2 does not need a temperature shift-up stage.

Of course, the iPSCs generated by those two different protocols exhibited the features of hESCs including the marker genes expression and *in vitro*/*in vivo* pluripotency (Ban et al. 2011).

7.4 Applications

7.4.1 Generation of iPSCs from Cord Blood CD34⁺ Cells

The SeV vector enabled highly efficient gene transfer into human cord blood (CB)-derived $CD34^+$ hematopoietic stem cells (Li et al. 2000; Jin et al. 2003). $CD34^+$ CB cells represent the earliest somatic stem cells and are expected to lack postnatal genomic aberrations caused by, for example, irritants from the environment and UV irradiation. These features unique to $CD34^+$ CB cells suggest that they are an ideal cell source for generating "gold standard iPSCs." From this aspect, the use of retroviral vectors for the generation of iPSCs from $CD34^+$ CB cells (Giorgetti et al. 2009; Takenaka et al. 2010) may be controversial because these vectors randomly integrate to the chromosomes of these cells.

On the other hand, using the SeV vector would be highly suitable to generate such desirable iPSCs because the vector does not interact with the chromosomes of target cells. We were able to obtain footprint-free CB-iPSCs by Protocol 1 using the TS7 SeV vectors (see Sect. 7.3.2 and Fig. 7.3a) (Ban et al. 2011). Interestingly, an MOI of only 2 was sufficient to generate iPSCs from human CB stem cells, possibly because of the high transduction efficiency of the vector for this cell type (Jin et al. 2003).

7.4.2 Generation of iPSCs from Peripheral Blood Cells, the Activated T Cells

Skin fibroblasts have been routinely used as the source of patient-specific iPSCs. These cells are obtained by invasive biopsy, which may scar patients. Moreover, a small number of cells obtainable by biopsy have to be extensively expanded *in vitro* before being used for iPSC generation, which takes at least 1 month and may facilitate the accumulation of unintended genetic mutations. In contrast, taking peripheral blood samples is far less invasive and easier to obtain informed consent from patients, compared with skin biopsies. Blood cells may be used after an incubation period of only several days for T-cell activation and expansion by an anti-CD3 antibody and IL-2 treatment.

Seki et al. (2010) demonstrated that vector/transgene-free human T-cell-derived iPSCs (TiPSCs) were generated from a small amount (less than 1 ml) of human peripheral blood within only 1 month using a combination of three human *OCT3/4*-, *SOX2*-, *KLF4*-carrying TS conv vectors and one c-*MYC*-carrying TS15 vector (Protocol 2, see Sect. 7.3.2 and Fig. 7.3c) at an MOI of 10–20. The efficiency of iPSC generation was about 0.1 %. It was much higher than that by the other methods using retroviral (Brown et al. 2010) or lentiviral (Loh et al. 2010; Staerk et al. 2010) vectors in which the efficiency was shown to be very low (approximately 0.0008–0.01 %) and 30–300 ml blood was required to initiate iPSC generation.

7.4.3 *Patient-Specific iPSCs for Better Understanding of Diseases and Development of Novel Therapeutic Options*

7.4.3.1 Establishment of a Patient-Specific iPSC Bank

Our collaboration with Kumamoto University, Kumamoto, Japan has enabled us to establish hundreds of human iPSC lines from the fibroblasts and blood samples of patients with various neural, metabolic, muscular, skin, bone, and other inherited intractable diseases. This work has been a part of the project of the Ministry of Health, Labor and Welfare, Japan for the establishment of a bank of vector/transgene-free patient-specific iPSC lines to advance research on difficult-to-treat diseases and drug screening. The footprint-free generation of iPSCs by SeV TS vectors is suitable for such a bank establishment because the method does not introduce any genetic alterations to the cells of patients, and also because the high efficiency and simplicity of the method are expected to facilitate the generation of iPSCs from various cell sources from patients. As a result of this project, a new drug candidate was found using transgene-free iPSCs derived from fibrodysplasia ossificans progressiva, a very rare inherited disease that causes unusual bone formation in the muscle (Hamasaki et al. 2012). TS vectors have also been used to generate footprint-free disease-specific iPSCs from elderly patients with type 1 and type 2 diabetes (Kudva et al. 2012). These iPSCs also will be useful for the disease modeling and screening of therapeutics for diabetes.

7.4.3.2 Regenerative Medicine with Targeted Gene Correction for Inherited Diseases

The severe shortage of donors and harmful effects of immunosuppressive treatments impose major limitations on organ transplantation, which makes the potential of human iPSC-based therapy highly attractive. AIATD (α1-antitrypsin deficiency) is an autosomal recessive disorder found in 1 of 2,000 individuals of Northern European descent and represents the most common inherited metabolic disease of the liver. It results from a single point mutation in the AIAT gene (the Z allele; Glu342Lys), which causes the protein to form ordered polymers within the endoplasmic reticulum of hepatocytes. The resulting inclusions cause liver cirrhosis, the only current therapy for which is liver transplantation.

Our collaboration with the Wellcome Trust Sanger Institute and University of Cambridge, UK, succeeded in generating AIATD patient-specific iPSCs by using a combination of three TS conv vectors carrying human *OCT3/4*, *SOX2*, and *KLF4*, respectively, and the c-*MYC*-carrying TS15 vector (Protocol 2, see Sect. 7.3.2 and Fig. 7.3c). The patient's specific iPSCs were amplified and subjected to an *in vitro* gene correction of the mutated sequences with zinc-finger nucleases and the PiggyBac technology (Yusa et al. 2011). The iPSCs with the corrected gene were then successfully differentiated into hepatocytes and transplanted to the liver of immunodeficient mice. These mice demonstrated that the transplanted patient's hepatocytes exhibited normal function in the liver. It should be noted again that the established iPSCs as well as the gene-corrected patient's hepatocytes were completely vector/transgene-free (footprint-free) and may keep the original genetic background of the patient with the only exception being the corrected AIAT gene. These results provide the first proof of principle for the potential of combining human iPSCs with genetic correction to generate clinically relevant transgene-free cells for autologous cell-based therapies.

7.5 Conclusions

As illustrated in other chapters, the nonintegrating SeV vector is a safe and efficient tool for gene transduction that can be used in very broad biological and medical settings. Here, we demonstrated that the use of ts genomes as the vector backbone allowed for the generation of transgene/vector-free iPSCs with high efficiency. Therefore, this technology appears to be able to resolve certain serious concerns in the field discussed here and is now being widely used. It also will be a useful tool for direct reprogramming to convert one type of somatic cell to another, but not via iPSCs. Switch-on/off of the expression of a transgene of interest as well as the removal of the transgene is often required in many other settings of cell biology and engineering. Thus, the application of ts SeV vectors is not limited to iPSC generation and direct conversion.

Acknowledgments We are grateful to M. Hasegawa and A. Iida at DNAVEC Corporation for critical reading of this manuscript. We thank Y. Yonemitsu at Kyushu University for his helpful advice.

References

Ban H, Inoue M, Griesenbach U, Munkonge F, Chan M, Iida A, Alton EW, Hasegawa M (2007) Expression and maturation of Sendai virus vector-derived CFTR protein: functional and biochemical evidence using a GFP-CFTR fusion protein. Gene Ther 14:1688–1694

Ban H, Nishishita N, Fusaki N, Tabata T, Saeki K, Shikamura M, Takada N, Inoue M, Hasegawa M, Kawamata S, Nishikawa S (2011) Efficient generation of transgene-free human induced pluripotent stem cells (iPSCs) by temperature-sensitive Sendai virus vectors. Proc Natl Acad Sci USA 108:14234–14239

Bowman MC, Smallwood S, Moyer SA (1999) Dissection of individual functions of the Sendai virus phosphoprotein in transcription. J Virol 73:6474–6483

Brown ME, Rondon E, Rajesh D, Mack A, Lewis R, Feng X, Zitur LJ, Learish RD, Nuwaysir EF (2010) Derivation of induced pluripotent stem cells from human peripheral blood T lymphocytes. PLoS One 5:e11373

Feller JA, Smallwood S, Skiadopoulos MH, Murphy BR, Moyer SA (2000) Comparison of identical temperature-sensitive mutations in the L polymerase proteins of Sendai and parainfluenza 3 viruses. Virology 276:190–201

Fusaki N, Ban H, Nishiyama A, Saeki K, Hasegawa M (2009) Efficient induction of transgene-free human pluripotent stem cells using a vector based on Sendai virus, an RNA virus that does not integrate into the host genome. Proc Jpn Acad Ser B Phys Biol Sci 85:348–362

Giorgetti A, Montserrat N, Aasen T, Gonzalez F, Rodríguez-Pizà I, Vassena R, Raya A, Boué S, Barrero MJ, Corbella BA, Torrabadella M, Veiga A, Izpisua Belmonte JC (2009) Generation of induced pluripotent stem cells from human cord blood using OCT4 and SOX2. Cell Stem Cell 5:353–357

Hamasaki M, Hashizume Y, Yamada Y, Katayama T, Hohjoh H, Fusaki N, Nakashima Y, Furuya H, Haga N, Takami Y, Era T (2012) Pathogenic mutation of ALK2 inhibits induced pluripotent stem cell reprogramming and maintenance: mechanisms of reprogramming and strategy for drug identification. Stem Cells 30:2437–2449

Inoue M, Tokusumi Y, Ban H, Kanaya T, Tokusumi T, Nagai Y, Iida A, Hasegawa M (2003) Nontransmissible virus-like particle formation by F-deficient Sendai virus is temperature sensitive and reduced by mutations in M and HN proteins. J Virol 77:3238–3246

Jia F, Wilson KD, Sun N, Gupta DM, Huang M, Li Z, Panetta NJ, Chen ZY, Robbins RC, Kay MA, Longaker MT, Wu JC (2010) A nonviral minicircle vector for deriving human iPS cells. Nat Methods 7:197–199

Jin CH, Kusuhara K, Yonemitsu Y, Nomura A, Okano S, Takeshita H, Hasegawa M, Sueishi K, Hara T (2003) Recombinant Sendai virus provides a highly efficient gene transfer into human cord blood-derived hematopoietic stem cells. Gene Ther 10:272–277

Kaji K, Norrby K, Paca A, Mileikovsky M, Mohseni P, Woltjen K (2009) Virus-free induction of pluripotency and subsequent excision of reprogramming factors. Nature (Lond) 458:771–775

Kim D, Kim CH, Moon JI, Chung YG, Chang MY, Han BS, Ko S, Yang E, Cha KY, Lanza R, Kim KS (2009) Generation of human induced pluripotent stem cells by direct delivery of reprogramming proteins. Cell Stem Cell 4:472–476

Kudva YC, Ohmine S, Greder LV, Dutton JR, Armstrong A, De Lamo JG, Khan YK, Thatava T, Hasegawa M, Fusaki N, Slack JM, Ikeda Y (2012) Transgene-free disease-specific induced pluripotent stem cells from patients with type 1 and type 2 diabetes. Stem Cells Transl Med 1:451–461

Li HO, Zhu YF, Asakawa M, Kuma H, Hirata T, Ueda Y, Lee YS, Fukumura M, Iida A, Kato A, Nagai Y, Hasegawa M (2000) A cytoplasmic RNA vector derived from nontransmissible Sendai virus with efficient gene transfer and expression. J Virol 74:6564–6569

Loh YH, Hartung O, Li H, Guo C, Sahalie JM, Manos PD, Urbach A, Heffner GC, Grskovic M, Vigneault F, Lensch MW, Park IH, Agarwal S, Church GM, Collins JJ, Irion S, Daley GQ (2010) Reprogramming of T cells from human peripheral blood. Cell Stem Cell 7:15–19

Nishimura K, Sano M, Ohtaka M, Furuta B, Umemura Y, Nakajima Y, Ikehara Y, Kobayashi T, Segawa H, Takayasu S, Sato H, Motomura K, Uchida E, Kanayasu-Toyoda T, Asashima M, Nakauchi H, Yamaguchi T, Nakanishi M (2011) Development of defective and persistent Sendai virus vector: a unique gene delivery/expression system ideal for cell reprogramming. J Biol Chem 286:4760–4771

Okita K, Ichisaka T, Yamanaka S (2007) Generation of germ-line-competent induced pluripotent stem cells. Nature (Lond) 448:313–317

Okita K, Nakagawa M, Hyenjong H, Ichisaka T, Yamanaka S (2008) Generation of mouse induced pluripotent stem cells without viral vectors. Science 322:949–953

Sakai Y, Kiyotani K, Fukumura M, Asakawa M, Kato A, Shioda T, Yoshida T, Tanaka A, Hasegawa M, Nagai Y (1999) Accommodation of foreign genes into the Sendai virus genome: sizes of inserted genes and viral replication. FEBS Lett 456:221–226

Seki T, Yuasa S, Oda M, Egashira T, Yae K, Kusumoto D, Nakata H, Tohyama S, Hashimoto H, Kodaira M, Okada Y, Seimiya H, Fusaki N, Hasegawa M, Fukuda K (2010) Generation of induced pluripotent stem cells from human terminally differentiated circulating T cells. Cell Stem Cell 7:11–14

Soldner F, Hockemeyer D, Beard C, Gao Q, Bell GW, Cook EG, Hargus G, Blak A, Cooper O, Mitalipova M, Isacson O, Jaenisch R (2009) Parkinson's disease patient-derived induced pluripotent stem cells free of viral reprogramming factors. Cell 136:964–977

Sommer CA, Stadtfeld M, Murphy GJ, Hochedlinger K, Kotton DN, Mostoslavsky G (2009) Induced pluripotent stem cell generation using a single lentiviral stem cell cassette. Stem Cells 27:543–549

Stadtfeld M, Nagaya M, Utikal J, Weir G, Hochedlinger K (2008) Induced pluripotent stem cells generated without viral integration. Science 322:945–949

Staerk J, Dawlaty MM, Gao Q, Maetzel D, Hanna J, Sommer CA, Mostoslavsky G, Jaenisch R (2010) Reprogramming of human peripheral blood cells to induced pluripotent stem cells. Cell Stem Cell 7:20–24

Takahashi K, Yamanaka S (2006) Induction of pluripotent stem cells from mouse embryonic and adult fibroblast cultures by defined factors. Cell 126:663–676

Takahashi K, Tanabe K, Ohnuki M, Narita M, Ichisaka T, Tomoda K, Yamanaka S (2007) Induction of pluripotent stem cells from adult human fibroblasts by defined factors. Cell 131:861–872

Takenaka C, Nishishita N, Takada N, Jakt LM, Kawamata S (2010) Effective generation of iPS cells from CD34+ cord blood cells by inhibition of p53. Exp Hematol 38:154–162

Tokusumi T, Iida A, Hirata T, Kato A, Nagai Y, Hasegawa M (2002) Recombinant Sendai viruses expressing different levels of a foreign reporter gene. Virus Res 86:33–38

Warren L, Manos PD, Ahfeldt T, Loh YH, Li H, Lau F, Ebina W, Mandal PK, Smith ZD, Meissner A, Daley GQ, Brack AS, Collins JJ, Cowan C, Schlaeger TM, Rossi DJ (2010) Highly efficient reprogramming to pluripotency and directed differentiation of human cells with synthetic modified mRNA. Cell Stem Cell 7:618–630

Woltjen K, Michael IP, Mohseni P, Desai R, Mileikovsky M, Hämäläinen R, Cowling R, Wang W, Liu P, Gertsenstein M, Kaji K, Sung HK, Nagy A (2009) PiggyBac transposition reprograms fibroblasts to induced pluripotent stem cells. Nature (Lond) 458:766–770

Yamanaka S (2010) Patient-specific pluripotent stem cells become even more accessible. Cell Stem Cell 7:1–2

Yoshizaki M, Hironaka T, Iwasaki H, Ban H, Tokusumi Y, Iida A, Nagai Y, Hasegawa M, Inoue M (2006) Naked Sendai virus vector lacking all of the envelope-related genes: reduced cytopathogenicity and immunogenicity. J Gene Med 8:1151–1159

Yu J, Hu K, Smuga-Otto K, Tian S, Stewart R, Slukvin II, Thomson JA (2009) Human induced pluripotent stem cells free of vector and transgene sequences. Science 324:797–801

Yusa K, Rashid ST, Strick-Marchand H, Varela I, Liu PQ, Paschon DE, Miranda E, Ordóñez A, Hannan NR, Rouhani FJ, Darche S, Alexander G, Marciniak SJ, Fusaki N, Hasegawa M, Holmes MC, Di Santo JP, Lomas DA, Bradley A, Vallier L (2011) Targeted gene correction of α1-antitrypsin deficiency in induced pluripotent stem cells. Nature (Lond) 478:391–394

Zhou H, Wu S, Joo JY, Zhu S, Han DW, Lin T, Trauger S, Bien G, Yao S, Zhu Y, Siuzdak G, Schöler HR, Duan L, Ding S (2009) Generation of induced pluripotent stem cells using recombinant proteins. Cell Stem Cell 4:381–384

Chapter 8
Gene Therapy for Peripheral Arterial Disease Using Sendai Virus Vector: From Preclinical Studies to the Phase I/IIa Clinical Trial

Yoshikazu Yonemitsu, Takuya Matsumoto, and Yoshihiko Maehara

Abstract Great potential for limb salvaging with basic fibroblast growth factor (FGF-2) expressed from a Sendai virus vector with the full-length viral genome (SeV/FGF-2) in the target muscle was demonstrated in animal models of severe limb ischemia. The exogenously expressed FGF-2 induced other endogenous angiogenic factors, including vascular endothelial growth factor and hepatocyte growth factor in a highly concerted fashion, and this was shown to underlie the remarkable limb salvaging. These results led to the initiation of a phase I/IIa clinical trial for peripheral arterial disease (PAD) of humans. To be better prepared for unforeseen adverse events, this clinical trial used the *F* gene-deleted, nontransmissible, and hence safer (than SeV/FGF-2) version that carried the human FGF-2 gene (ΔFSeV/FGF-2; product code, DVC1-0101). Overall, DVC1-0101 appeared to be safe, giving rise to no serious adverse events in various criteria. Moreover, it exerted a significant therapeutic effect from a number of clinical aspects. DVC1-0101 represents the first case of the potential diverse medical applications of SeV vector. It is now hoped to move on to an advanced phase of clinical trial with a larger number of patients and placebo group to firmly establish the safety and efficacy of DVC1-0101.

Y. Yonemitsu (✉)
R&D Laboratory for Innovative Biotherapeutics, Kyushu University Graduate School of Pharmaceutical Sciences, Fukuoka 812-8582, Japan
e-mail: yonemitu@med.kyushu-u.ac.jp

T. Matsumoto • Y. Maehara
Department of Surgery and Science, Kyushu University Graduate School of Medical Sciences, Fukuoka 812-8582, Japan

Y. Nagai (ed.), *Sendai Virus Vector: Advantages and Applications*,
DOI 10.1007/978-4-431-54556-9_8, © Springer Japan 2013

8.1 Introduction

Peripheral arterial disease (PAD) is a typical phenotype of progressive and systemic atherosclerosis that causes disability of limb function (IC, intermittent claudication) as well as serious pain or limb loss (CLI, critical limb ischemia). PAD is estimated to affect more than 8.4 million individuals in the United States (Weitz et al. 1996), and its prevalence is increasing in Asian countries including Japan. With the current standard therapies, subjects with severe PAD are left with immobility, intractable ischemia, ulceration, impaired wound healing, or the necessity of amputation. Therefore, the development of new therapeutics to rescue the diseased limbs, normalize limb function, and prevent the progression of PAD is urgently needed.

In contrast to interventional revascularization and bypass formation that target the large and medium-sized arteries, therapeutic angiogenesis modalities target microvessels (Takeshita et al. 1994). Soon after the discovery of angiogenic growth factors, researchers began to test those protein factors (Lazarous et al. 2000; Lederman et al. 2002) and the genes encoding them (Comerota et al. 2002; Rajagopalan et al. 2003) for their potential to stimulate the formation of neo-vessels in subjects with PAD. So far, however, the research outcomes from the clinical trials for therapeutic angiogenesis have been unsatisfactory since the positive findings sometimes obtained in the small early trials have not been confirmed in the larger, well-controlled multicenter trials.

Among the published angiogenic trials for IC, only the TRAFFIC study, a well-controlled clinical trial with the intraarterial injection of human recombinant basic fibroblast growth factor (bFGF/FGF-2), provided the proof-of-concept; namely, a single-dose injection of FGF-2 significantly improved the peak walking time of IC patients at day 90 (Lederman et al. 2002). However, the TRAFFIC trial was terminated because of the relatively limited clinical outcomes and the mild-to-moderate adverse events (AEs), including hypotension and proteinuria. The point could be that systemic delivery of FGF-2 via intraarterial injection was adopted in this trial. It probably did not achieve the local concentration sufficient to produce satisfactory therapeutic outcomes on one hand and on the other exerted systemic influence to cause AEs of various kinds Therefore, to gain higher efficacy and reduce AEs, an approach is needed that will give rise to a higher and sustained concentration of FGF-2 in the local target muscles without systemic leakage.

Because a number of preclinical evaluations in animal studies supported the concept of therapeutic angiogenesis, the efforts should be continued in preclinical studies to identify critical factors that confer sufficient efficacy and then be extended to clinical studies.

Recombinant Sendai virus vector (rSeV) is a powerful gene transfer agent for the cytoplasmic expression of therapeutic genes (Yonemitsu et al. 2000; also see Chap. 3) We previously developed mouse models of surgically induced severe limb ischemia and demonstrated that rSeV expressing FGF-2 exhibited great therapeutic potential for preventing the auto-amputation of diseased limbs following administration into the local muscles (Masaki et al. 2002; Shoji et al. 2003). In parallel, the rSeV-expressing vascular endothelial growth factor (VEGF) failed to show such a high therapeutic potential.

These results and subsequent studies demonstrated that the therapeutic effect of FGF-2 was not attributable simply to its high level of local expression alone. Rather, triggering by the exogenously expressed FGF-2 of the angiogenic cascade to induce other endogenous angiogenic factors such as VEGF and hepatocyte growth factor (HGF) in a well-harmonized fashion was critically important for the therapeutic potential (Masaki et al. 2002; Onimaru et al. 2002; Tsutsumi et al. 2004; Fujii et al. 2006).

Overall, the preclinical studies, along with the availability of a safer vector version (see below), provided the biological and pharmaceutical rationale for moving on to a clinical trial.

In the preclinical studies we used the SeV vector with the full-length viral genome (the first-generation SeV vector). In the clinical trial, the *F* gene-deleted nontransmissible Δ*F*SeV vector (the second-generation vector) was used to maximize safety without much attenuating the expression capacity (see Chap. 3). The construct Δ*F*SeV/FGF-2 was coded for DVC1-0101.

This chapter briefly describes the research outcomes in the preclinical studies and then outlines the procedures and the results regarding safety and efficacy of the DVC1-0101 phase I/IIa trial.

8.2 Preclinical Studies at the Animal Level

8.2.1 Establishment of a Severe Acute Murine Auto-Amputation Model of Limb Ischemia

Five-week-old male balb/c nu/nu mice were used to develop a murine "auto-amputation model" in which the mice consistently lost their limbs during the course of the 10-day experiment (for experimental details, see Masaki et al. 2002). Briefly, the entire left saphenous artery and vein, and the left external iliac artery and vein with deep femoral and circumflex arteries and veins, were ligated, cut, and excised to set up a murine model of "severe hind limb ischemia." For gene transfer, 25 μl SeV/FGF-2 solutions were injected into two sites of the local thigh muscle soon after completion of the surgical procedures. The limb prognosis was expressed according to the limb salvage index as described (Masaki et al. 2002).

8.2.2 SeV Vector Construction

There are several sites throughout the SeV genome into which a foreign gene of interest can be inserted (see Chap. 3). The human FGF-2 ORF was inserted to the *le/N* site along with the SeV transcription unit in the full-length SeV genome cDNA, the first generation of the SeV vector. Insertion to the *le/N* site allowed the highest level of expression compared with insertions into the various downstream sites (see Chap. 3). This expression vector was named SeV/FGF-2. SeV-expressing VEGF (SeV/VGEF165) was also constructed similarly.

8.2.3 Demonstration of the Therapeutic Potential of SeV/FGF-2 Using the Mouse Model

Masaki et al. (2002) showed significant inhibition of the auto-amputation by SeV/FGF-2, whereas expression of VEGF from SeV/VEGF165 rather accelerated limb loss. We examined the vector-dose dependency of FGF-2 expression in the local muscles and the therapeutic outcome. The net FGF-2 (sum of the exogenously expressed human FGF-2 and the endogenously expressed basal level of murine FGF-2) in the muscles increased in amount proportionally to the titer of input vector (Fig. 8.1a). However, the therapeutic outcomes displayed no such correlation. All the doses tested exhibited therapeutic potential, but doses around 3×10^6 to 1×10^7 PFU exhibited the highest efficacy (Fig. 8.1b). For input doses lower or higher than these, the outcomes were less remarkable, indicating that there could be an optimal dose. In addition, toxicity might even develop because when a far excess dose (1×10^8 PFU) was used that produced more than 300 ng/g muscle protein FGF-2, the therapeutic effect was lost altogether (Masaki et al. 2002). Therefore, it is important to optimize the input dose, which was one of the reasons the phase I/IIa trial adopted a dose-escalation scheme (see Sect. 2).

8.2.4 Molecular Bases Underlying the Therapeutic Effect of FGF-2

Extensive studies were conducted to understand the mechanism responsible for limb salvage by FGF-2 expressed from SeV vector. One of the key results obtained was that SeV-based FGF-2 expression stimulated the synthesis of the endogenous angiogenic factors VEGF and HGF at an early phase in the angiogenic cascade (Masaki et al. 2002; Onimaru et al. 2002). The requirement of induction of endogenous factors for the therapeutic effect of FGF-2 was strongly suggested by the fact that neutralization of either VEGF or HGF by the respective specific antibodies *in vivo* abolished the therapeutic potential of FGF-2 (Masaki et al. 2002; Onimaru et al. 2002). This induction of endogenous VEGF and HGF appeared to take place via a platelet-derived growth factor receptor-α/p70S6-mediated mechanism (Tsutsumi et al. 2004). FGF-2 gene expression from the SeV vector induced not only angiogenesis but also lymphangiogenesis via the VEGF-C/VEGFR3 system (Kubo et al. 2002; Onimaru et al. 2009). FGF-2 also stimulated transient

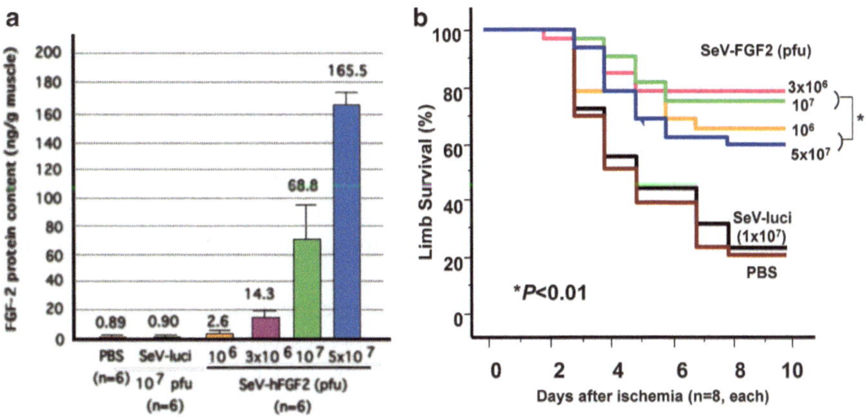

Fig. 8.1 Expression of fibroblast growth factor (FGF-2) (**a**) and limb prognosis curve (**b**) in ischemic thigh muscles of the murine auto-amputation model after Sendai virus (SeV)-mediated FGF-2 gene transfer, indicating the possible relationship between levels of transgene expression and therapeutic effect. $p<0.05$, $*p<0.01$. (**a**) *Bar graph* indicates dose-dependent increase of net expression of FGF-2, including endogenous and exogenous FGF-2. Two days after operation and vector injection, a posterior portion of thigh muscles was subjected to ELISA, which cross-reacts to both murine and human FGF-2. Each group contains six animals. Data are expressed as mean ± SEM. (**b**) Limb prognosis curve following intramuscular injection of phosphate-buffered saline (PBS), control vector (SeV-luciferase), or SeV-hFGF2. Each group included eight mice with an auto-amputation model. Curve was obtained using the Kaplan–Meier method. Data were analyzed using log-rank test. (Adapted from Shoji et al. 2003)

inflammatory/arteriogenic responses via a monocyte/macrophage chemoattractant protein-1 (MCP-1)-mediated mechanism (Fujii et al. 2006).

Taken together, the preclinical studies indicated that FGF-2, which has the potential to induce other angiogenic factors including VEGF and HGF and triggers other biological events, is likely a major conductor of the "angiogenic orchestra" and that limb salvage is not possible without such fine orchestration. Overall, our results together with the clinical information derived from the TRAFFIC study strongly suggested that the intramuscular injection of SeV/FGF-2 would represent an effective method for treating PAD.

8.3 The Phase I/IIa Open-Label, Dose-Escalation Clinical Trial

8.3.1 Use of a Nontransmissible Safer Vector Version

The primary endpoint in a phase I/IIa trial is to assess safety. No link between SeV and any human diseases has been reported over the several decades since the discovery of the virus (see Chap. 2). Although its receptor is the terminal sialic acid residues that are ubiquitously expressed throughout the body, SeV causes a localized infection but not a systemic infection because the proteases essential for cleavage-activation of the viral fusion glycoprotein and hence spreading of the virus by multiple cycles of replication are expressed by limited tissue types (see Chaps. 2 and 3). Therefore, the SeV vector in the backbone of the full-length viral genome appeared to be safe enough for humans. However, the slightest possibility still remained that the injected SeV vector would spread within the target body and further be transmitted to nearby persons so long as a vector with the full-length viral genome is used.

Therefore, attempts have been made to generate theoretically nontransmissible vector versions by deleting one or more essential genes and recovering such viruses from cells expressing the respective genes in *trans* (see Chap. 3). Of a series of the deletion mutants generated, that *F* gene deleted (Δ*F*SeV) was able to be recovered with the least difficulty and reached the highest titers, and hence appeared to be of the highest practical merit. This version was thus used, and the gene encoding the human FGF-2 was inserted into the *le/N* site (see Chap. 3). The product Δ*F*SeV/hFGF2 was given the product code, DVC1-0101. The safety and tolerability as well as possible therapeutic effects were assessed following a single administration of DVC1-0101 into the local target muscle of patients with PAD.

The phase I/IIa clinical trial of DVC1-0101 was approved and initiated in January 2006 and was terminated in March 2011. In the following sections, we outline the procedures and outcomes of this trial according to the recently published details (Yonemitsu et al. 2013).

8.3.2 Baseline Patient Characteristics

The Japanese Guidelines for Clinical Trials of Gene Therapy issued by the Ministry of Health, Labor and Welfare (MHLW) stipulate that only "no-option" patients should be enrolled as the subjects in the first-in-human gene therapy studies. Therefore, the governmental review board recommended that this trial would target the cases of CLI without indications for any standard vascular interventions. In addition, because DVC1-0101 was designed to improve the walking performance of PAD patients, we made an effort to recruit subjects who could undergo a treadmill test.

A total of 18 patients with PAD who met the inclusion criteria were screened, and among them, 12 patients who did not meet the exclusion criteria were enrolled (Table 8.1). A total of 12 limbs, 1 limb per patient, were treated once with DVC1-0101 by injecting into multiple sites of the target muscles in a four-dose escalation fashion from stage 1 to stage 4. Each stage enrolled 3 patients. The doses (cell infectious units, CIU/60 kg) used were 5×10^7 (stage 1), 2×10^8 (stage 2), 1×10^9 (stage 3), and 5×10^9 (stage 4). The entire cohort consisted of 10 men and 2 women (mean age, 65.0 years; range, 48–82), including 10 cases of arteriosclerosis obliterans (ASO) and 2 of thromboangitis obliterans (TAO, Buerger's disease; both cases were in stage 4). Three patients had single (cases 103 and 403) or multiple (case 105) ischemic ulcers, and 8 patients had relatively mild-to-moderate rest pain that did not prevent them from taking a treadmill test. Nine patients had a smoking history, and enrollment was restricted to the patients who had ceased smoking at least 1 month before screening. For more details, see Table 8.1.

8.3.3 Primary Endpoint: Safety

8.3.3.1 Survival, Adverse Events (AEs), and Virus Shedding

Although one patient (case 105, stage 1) was lost from acute-on-chronic progression of preexisting interstitial pneumonia approximately 2 years after treatment, no other death occurred during a 1-year period (the observation period ranged from 1 year and 1 month to 4 years and 8 months).

In the clinical protocol used, serious adverse events (SAEs) are to be reported to and reviewed by the Data Safety and Monitoring Board, and followed by the Health Sciences Council of MHLW of Japan for 5 years after the end of the protocol. Any other adverse events are to be carefully followed for 6 months after the gene transfer. Nine SAEs occurred in six patients during the observation period. The reviewing committee concluded that eight of them were unrelated to the gene transfer, although the relationship could not be denied for one (Yonemitsu et al. 2013).

A total of 136 AEs were observed (pretreatment = 16 events; posttreatment = 120 events) during the 6 months after the gene transfer. Among these there was no AE categorized in grade 4. Seven of the eight AEs related to amputation made in stage 1 were grade 3 according to the Common Terminology Criteria for Adverse Events (CTCAE version 4.03, published by the U.S. National Cancer Institute, 14 June, 2010: http://evs.nci.gov/ftp1/CTCAE/CTCAE_4.03_2010-06-14_QuickReference_8.5x11.pdf) Another AE at stage 4 was the result of alcoholic liver damage. No dose–response relationship was observed.

Urine and whole blood samples were collected from all the patients and used to monitor the virus shedding in individuals by measuring the levels of genome copies using nested real-time reverse transcriptase-polymerase chain reaction (RT-PCR) and virus amounts by the chicken erythrocytes agglutination activity (hemagglutination, HA). Neither the viral RNA genome nor HA activity was detected in urine samples. Three patients (cases 303, 401, and 405) exhibited transient presence of the viral genome in the blood. No HA (virus particle production) was found in any samples, including those that tested positive for the RNA genome.

8.3.3.2 Proinflammatory Reactions

All 12 patients were administered methylprednisolone hemisuccinate (Solu-medrol; 125 mg/day) via intravenous drip on day 0 (just before the vector injection) and day 1 (after vector injection) to avoid an unexpected evocation of the innate immune

Table 8.1 Patients' baseline characteristics ($n = 12$)

					Clinical characteristics					
Cohort	Case no.	Age (years)	Gender	Smoking history	Target limb	Diagnosis[a]	Rest pain grade[b]	Ulcer diameter (mm)	Available for treadmill	Fontaine/ Rutherford
Stage 1	102	59	M	+	Right	ASO	2	–	Yes	III/II-4
	103	74	M	+	Left	ASO	4	23.1 (5th toe) 11.3 (ankle) 9.5 (foot sole)	No	IV/III-5
	105	66	M	–	Left	ASO	4	8.4 (5th toe)	No	IV/III-5
Stage 2	201	65	F	–	Right	ASO	2	–	Yes	III/II-4
	203	58	M	+	Left	ASO	2	–	Yes	III/II-4
	204	76	F	–	Right	ASO	2	–	No[c]	III/II-4
Stage 3	303	82	M	+	Right	ASO	2	–	Yes	III/II-4
	304	58	M	+	Right	ASO[d]	3	–	Yes	III/II-4
	305	83	M	+	Left	ASO	2	–	Yes[e]	III/II-4
Stage 4	401	48	M	+	Right	TAO	2	–	Yes	III/II-4
	403	54	M	+	Right	TAO	3	8.0 (4th toe)	Yes	IV/III-5
	405	57	M	+	Right	ASO	4	–	No[f]	III/II-4

[a] *ASO* arteriosclerosis obliterans, *TAO* thromboangitis obliterans
[b] *1* no pain without analgesic, *2* tolerable pain without analgesic, *3* tolerable with a few doses of analgesic(s), *4* multiple doses of analgesic(s), *5* insufficient pain control with constant use of analgesic(s)
[c] Available for treadmill after DVC1-0101 treatment
[d] A case of high aortic occlusion just below renal arteries
[e] Not available because of breathlessness 2 months after DVC1-0101 treatment
[f] Preexisting hemiplasia caused by old cerebral infarction

response and allergic reactions to the vector. Administration of this cortisone has been shown to suppress the innate immune response but not antigen-specific acquired immune responses against SeV vectors under experimental conditions (Ikeda et al. 2002). So far, nothing particular has been recognized regarding the patients' laboratory parameters of systemic inflammatory reactions related to SeV.

In most of the patients, the white blood cell count (WBC) increased transiently soon after gene transfer and returned to baseline by day 3. Because no corresponding increase of C-reactive protein (CRP) or other proinflammatory cytokines was found, the increase of WBC counts was thought to be a result of the methylprednisolone treatment. In contrast, the levels of CRP and interleukin (IL)-6 were increased on days 7–14 on average. However, the averaged increase was largely attributable to the marked elevation of these parameters in a special case, case 103, who underwent a major amputation after showing a septic reaction to a methicillin-resistant *Staphylococcus aureus* infection. Other proinflammatory cytokines, including interleukin-1β and tumor necrosis factor-α, did not show a significant change in amount during the trial. Thus, so far no serious inflammatory responses appeared to take place during the experimental period.

8.3.3.3 Circulating Angiogenic Factors

To monitor the possible leakage into the bloodstream of exogenously expressed FGF-2 and FGF-2-induced endogenous downstream angiogenic factors, VEGF and HGF (Tsutsumi et al. 2004), we subjected the plasma and serum samples to analysis with specific enzyme-linked immunosorbent assay (ELISA) systems. We used samples within 7 days of their collection because our previous studies demonstrated that the expression of exogenous gene products peaked at 2 days and declined at around 7 to 14 days after gene transfer using multiple genes and vector constructs (Masaki et al. 2002; Shoji et al. 2003). It was found that there was no significant elevation of any angiogenic factors examined in the plasma and sera, indicating little or no leakage of those factors into the blood during the observation period.

8.3.3.4 Other Examinations Related to Safety

All patients underwent a retinal examination by independent ophthalmologists, cancer screening with tumor markers, and computed tomography (CT) scans of the brain and whole body as part of a follow-up examination for 6 months after the gene transfer. No newly developed lesions were found in any examination.

8.3.4 Secondary Endpoint: Efficacy-Related Parameters

Because CLI patients without indications for standard vascular interventions were eligible for this trial, we monitored the time-course of clinical symptoms

(Rutherford Classification, rest pain, and ulcer healing) (Rutherford et al. 1997) and other surrogates possibly related to the hemodynamics, including ankle-brachial pressure index (ABPI) and toe pressure index (TPI), laser Doppler perfusion images (LDPIs), foot-pad temperature assessed by thermography, and pulse-volume recording (PVR), for each patient visit for 6 months after the gene transfer. In addition, a flat treadmill test at 2.4 km/h, terminated at 300 m to minimize the risk for cardiovascular complications, was added for the eight patients who were able to undergo the test ($n=6$ for ASO and $n=2$ for TAO; see Table 8.1). We summarize the results next. For details, see Yonemitsu et al. (2013).

8.3.4.1 Improvement of Clinical Staging (Rutherford Classification)

All CLI patients enrolled in this trial were categorized in Rutherford grade II-4 or III-5. At the visit 6 months after the gene transfer, each patient's Rutherford category was reevaluated and scored as ±1 (improved/worsened by one category), ±2 (improved/worsened by two categories) and ±3 (improved/worsened by more than three categories). The results were favorable: the overall improvement rate was 58.3 % ($n=7/12$, $p=0.7744$), and the improvement for stages 2–4 was 77.8 % ($n=7/9$, $p=0.1797$).

8.3.4.2 Rest Pain and Ischemic Ulcer

Rest pain was scored based on the frequency of intake of analgesics, in most cases nonsteroidal antiinflammatory drugs (NSAIDs), within 24 h of the interview, and was categorized in (1) completely pain free without analgesics, (2) feel pain but no need for analgesics, (3) feel pain and sometimes need analgesics, (4) need analgesics constantly to control the pain, and (5) uncontrollable pain despite continuous use of analgesics. Two patients (cases 103 and 105) were excluded because of major and minor amputation, respectively.

Overall ($n=10$), significant pain reduction was seen and continued for more than 6 months ($p<0.01$ or $p<0.05$, Wilcoxon's signed-rank test), and six patients were completely pain free during the observation period. The time-course of a preexisting ischemic ulcer was recorded in three patients (cases 103 and 105 at stage 1, and case 403 at stage 4). No improvement in size of the referenced ulcers was observed in the two stage 1 patients, and complete healing was observed in one stage 4 patient.

8.3.4.3 ABPI and TPI

To avoid examiner-dependent biases, ABPI and TBI were measured by an automatic oscillometric system. When the blood pressure of the target limb was too low to be detected, the theoretical lower limit, 10 mmHg, was used for calculation. No significant change was observed at any time point up to 6 months after the gene transfer.

8.3.4.4 LDPI and Foot-Pad Temperature

The blood flow ratio of both legs of each patient was measured using a LDPI analyzer. To minimize data variables resulting from ambient light and temperature, the LDPI index was expressed as the ratio of the blood flow in the treated limb to that in the contralateral limb. No significant change was observed as of 1 month after gene transfer. The foot skin temperature of both limbs was measured by independent and blinded physicians (dermatologist at the hospital). The measurements were performed at constant room temperature (25 °C), and a 60-min interval was allocated for the patient's acclimatization after the removal of his or her socks and shoes. A significant decrease in the differences of foot-pad temperature (°C = treated limb–untreated limb) was observed at 1 month after the gene transfer ($p < 0.05$, paired t test).

8.3.4.5 Pulse-Volume Recording (PVR)

The ten patients, excluding the two amputated cases, underwent constant PVR measurements. Seven patients (cases 102, 201, 204, 304, 305, 403, 405) showed negative PVR values at pretreatment, and six cases (excluding case 102) occasionally exhibited significant appearance of PVR values after the gene transfer.

8.3.4.6 Walking Ability Assessed by Treadmill

Eight patients ($n = 6$ for ASO and $n = 2$ for TAO) were eligible for the flat treadmill test; the test was performed at 2.4 km/h and terminated at 300 m to minimize the risk of cardiovascular complications. Regarding the absolute claudication distance (ACD)-based walking performance, all six of the patients who did not have TAO demonstrated steady and significant improvement (Fig. 8.2a). Increases of ACD (ΔACD, %) were also observed in five of these patients (one patient was excluded because of the lack of pretreatment data as the result of very severe rest pain) (Fig. 8.2b). Even when the data included the two cases of Buerger's disease (TAO), the trends of ACD and ΔACD were not influenced, suggesting that DVC1-0101 therapy would significantly improve the walking performance in patients with PAD.

8.3.5 A Case Report

Figure 8.3 presents the findings for case 201, a 65-year-old Japanese woman at stage 2 with both legs affected and with approximately 200 m of intermittent claudication since 1993 as a result of PAD. Her symptoms had been stable with beraprost and cilostazol treatment for the past 14 years; she then developed rest pain in the right foot, which was treated with nonsteroidal antiinflammatory drugs. Her angiogram demonstrated complete obstruction of the whole superficial femoral to

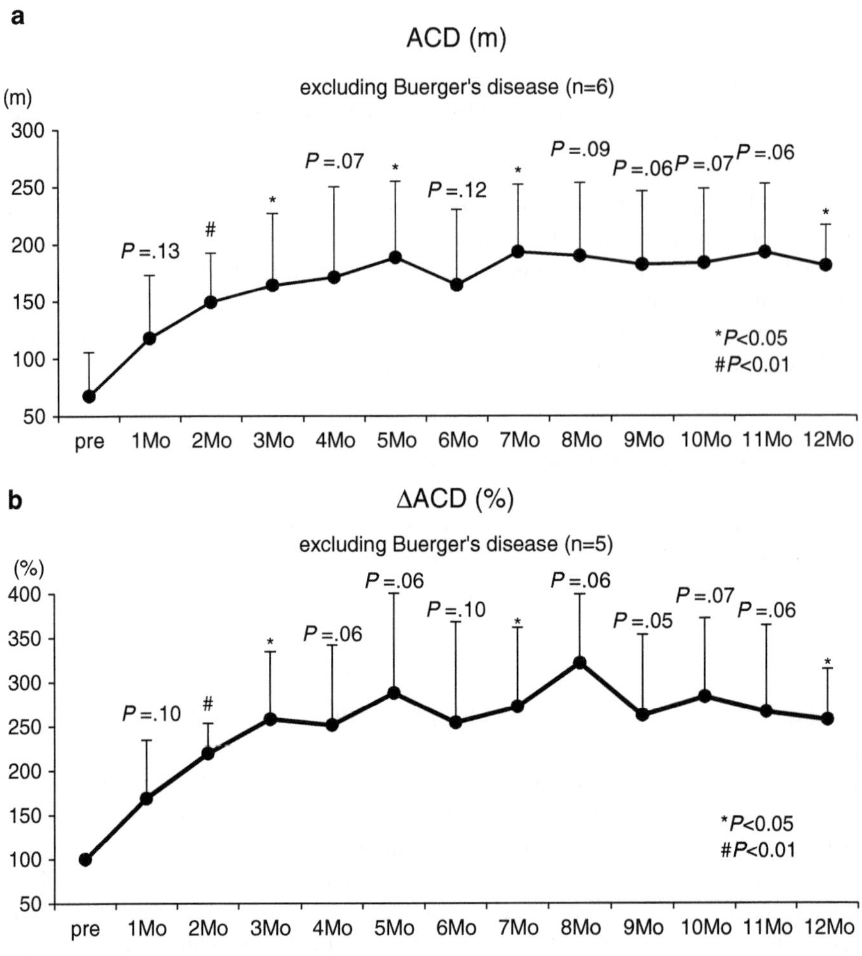

Fig. 8.2 (**a**) Time-course of absolute claudication distance (*ACD*: m, $n=6$). (**b**) Change of ACD (*ΔACD*: %, $n=5$). *$p<0.01$, #$p<0.05$. Data for two patients with Buerger's disease were excluded. (Adapted from Yonemitsu et al. 2013)

popliteal arteries and all main arteries at the ankle level in both limbs (Fig. 8.3a, left). Severe calcification throughout the superficial femoral to popliteal arteries on soft X-ray (Fig. 8.3a, right, white arrows) and insufficient size of both saphenous veins suggested the difficulty of surgical management; therefore, she was enrolled in the present trial on October 22, 2007.

She received DVC1-0101 at 30 sites of the right leg (10 sites at the upper and lower thigh and calf, respectively) with a total of 1.63×10^8 cell infection units (CIU). Her experimental course was uneventful. A thermographic examination demonstrated a right leg-specific increase of foot temperature (Fig. 8.3b, white arrows), and a PVR showed right first toe-specific pulsation as well (Fig. 8.3c). Her rest pain completely disappeared within 1 week after the gene transfer, and her

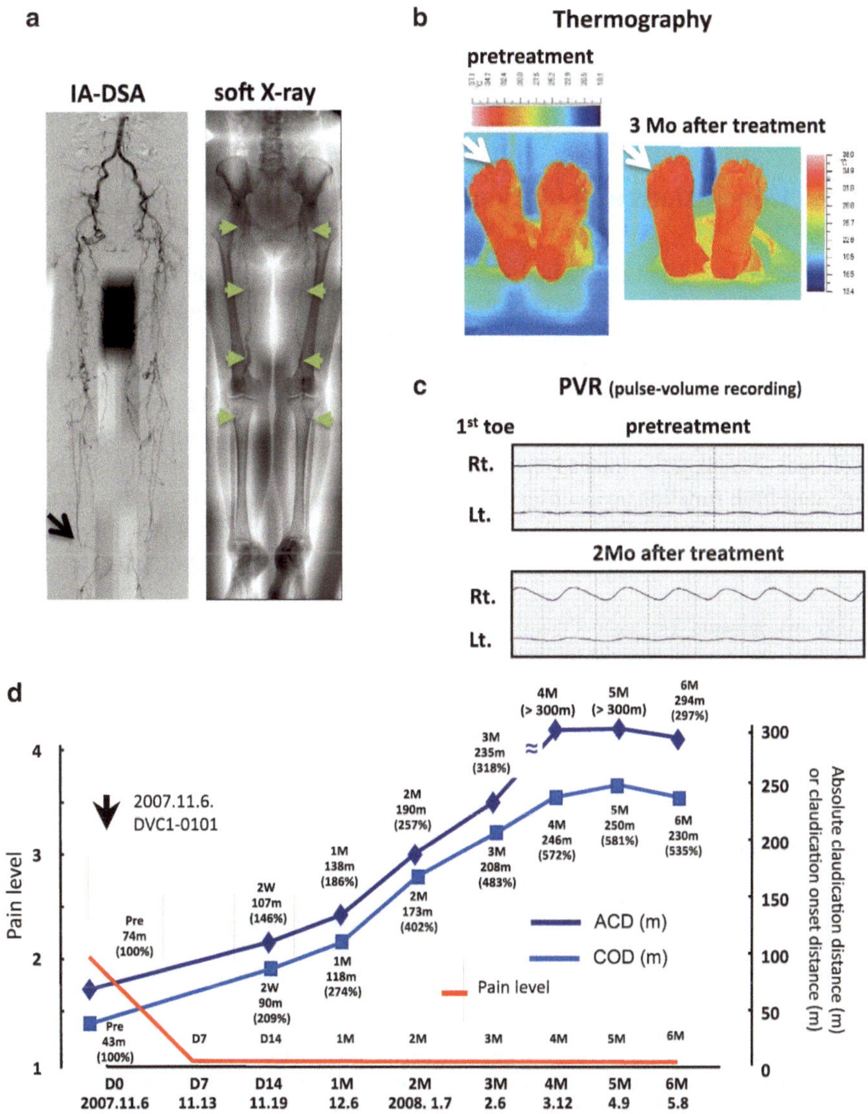

Fig. 8.3 A 65-year-old Japanese woman (case no. 201 at stage 2). Both limbs were affected, with approximately 200 m of intermittent claudication since 1993 caused by peripheral artery disease (PAD). Her symptoms had been stable for 14 years; she then developed rest pain in the right foot, which was treated by NSAIDs. (**a**) Her angiogram (*left panel*) revealed complete obstruction of the whole superficial femoral to popliteal arteries and all main arteries at the ankle level (*black arrow*) in both limbs. Severe calcification was seen throughout the superficial femoral to popliteal arteries on soft X-ray (*right panel*, *white arrows*). (**b**) Thermographic examination showed right limb-specific increase of foot temperature (*white arrows*) at 3 months after gene transfer. (**c**) A pulse-volume recording (PVR) showed right first toe-specific pulsation. (**d**) Clinical course. Her rest pain completely disappeared within 1 week after the gene transfer, and her treadmill examination showed linear increase of absolute claudication distance (*ACD*) and claudication onset distance (*COD*) at 4 months after the gene transfer. These improvements in her clinical symptoms have been maintained at 5 years after the gene transfer. *Mo*, months. (Adapted from Yonemitsu et al. 2013)

treadmill examination demonstrated a linear increase of ACD and claudication onset distance (COD) within 4 months after the gene transfer (Fig 8.2d). These improvements of clinical symptoms have been maintained for nearly 5 years thereafter.

8.4 Conclusions and Outlook

In the animal model of severe limb ischemia, FGF-2 expressed in the local target muscle from SeV vector was found to be highly beneficial for limb salvaging. This achievement appeared to be based upon the finely orchestrated expression of other endogenous angiogenic factors such as VEGF and HGF, which was triggered by the exogenously expressed FGF-2. This preclinical study moved on to a phase I/IIa clinical study that enrolled 12 PAD patients and used DVC1-0101 (Δ*F*SeV/hFGF-2) in a dose-escalation scheme. Overall, no serious adverse events took place. Moreover, significant improvement of limb functions was observed. Thus keenly awaited now is to advance to the next stage of clinical trial with a larger number of patients and a placebo control, which is expected to firmly establish the safety and efficacy of DVC1-0101.

Acknowledgments The authors thank Professor Yoichi Nakanishi, Mss. Keiko Kikutake, Michi Onizuka, and Nozomi Kataoka for their clinical research coordination, Ms. Akiko Kanaya for her help with drug preservation, Ms. Mio Yoneda-Maehara for her help with the trial paperwork, and Ms. Rio Yamauchi, Mr. Takeshi Maruoka, and Minoru Ido at EPS Co. for their help with the trial management. This study was supported in part by a Grant for Translational Research from MHLW to Y.Y.

References

Comerota AJ, Throm RC, Miller K, Henry T, Chronos N, Laird J, Sequeira R, Kent CK, Bacchetta M, Goldman C, Salenius JP, Schmieder FA, Pilsudski R (2002) Naked plasmid DNA encoding fibroblast growth factor type 1 for the treatment of end-stage unreconstructible lower extremity ischemia: preliminary results of a phase I trial. J Vasc Surg 35:930–936

Fujii T, Yonemitsu Y, Onimaru M, Tanii M, Nakano T, Egashira K, Takehara T, Inoue M, Hasegawa M, Kuwano H, Sueishi K (2006) Nonendothelial mesenchymal cell-derived MCP-1 is required for FGF-2-mediated therapeutic neovascularization: critical role of the inflammatory/arteriogenic pathway. Arterioscler Thromb Vasc Biol 226:2483–2489

Ikeda Y, Yonemitsu Y, Sakamoto T, Ishibashi T, Ueno H, Kato A, Nagai Y, Fukumura M, Inomata H, Hasegawa M, Sueishi K (2002) Recombinant Sendai virus-mediated gene transfer into adult rat retinal tissue: efficient gene transfer by brief exposure. Exp Eye Res 75:39–48

Kubo H, Cao R, Brakenhielm E, Mäkinen T, Cao Y, Alitalo K (2002) Blockade of vascular endothelial growth factor receptor-3 signaling inhibits fibroblast growth factor-2-induced lymphangiogenesis in mouse cornea. Proc Natl Acad Sci USA 99:8868–8873

Lazarous DF, Unger EF, Epstein SE, Stine A, Arevalo JL, Chew EY, Quyyumi AA (2000) Basic fibroblast growth factor in patients with intermittent claudication: results of a phase I trial. J Am Coll Cardiol 36:1239–1244

Lederman RJ, Mendelsohn FO, Anderson RD, Saucedo JF, Tenaglia AN, Hermiller JB, Hillegass WB, Rocha-Singh K, Moon TE, Whitehouse MJ, Annex BH, TRAFFIC Investigators (2002) Therapeutic angiogenesis with recombinant fibroblast growth factor-2 for intermittent claudication (the TRAFFIC study): a randomised trial. Lancet 35:2053–2058

Masaki I, Yonemitsu Y, Yamashita A, Sata S, Tanii M, Komori K, Nakagawa K, Hou X, Nagai Y, Hasegawa M, Sugimachi K, Sueishi K (2002) Angiogenic gene therapy for experimental critical limb ischemia: acceleration of limb loss by overexpression of VEGF165 but not of FGF-2. Circ Res 90:966–973

Onimaru M, Yonemitsu Y, Tanii M, Nakagawa K, Masaki I, Okano S, Ishibashi H, Shirasuna K, Hasegawa M, Sueishi K (2002) FGF-2 gene transfer can stimulate HGF irrespective of hypoxia-mediated downregulation in ischemic limbs. Circ Res 91:923–930

Onimaru M, Yonemitsu Y, Fujii T, Tanii M, Nakano T, Nakagawa K et al (2009) VEGF-C regulates lymphangiogenesis and capillary stability by regulation of PDGF-B. Am J Physiol Heart Circ Physiol 297:H1685–H1696

Rajagopalan S, Mohler ER III, Lederman RJ, Mendelsohn FO, Saucedo JF, Goldman CK, Blebea J, Macko J, Kessler PD, Rasmussen HS, Annex BH (2003) Regional angiogenesis with vascular endothelial growth factor in peripheral arterial disease: a phase II randomized, double-blind, controlled study of adenoviral delivery of vascular endothelial growth factor 121 in patients with disabling intermittent claudication. Circulation 108:1933–1938

Rutherford RB, Baker JD, Ernst C, Johnston KW, Porter JM, Ahn S, Jones DN (1997) Recommended standards for reports dealing with lower extremity ischemia: revised version. J Vasc Surg 26:517–538

Shoji T, Yonemitsu Y, Komori K, Tanii M, Itoh H, Sata S, Shimokawa H, Hasegawa M, Sueishi K, Maehara Y (2003) Intramuscular gene transfer of FGF-2 attenuates regenerative endothelial dysfunction and inhibits neointimal hyperplasia of autologous femoral vein grafts in poor runoff limbs of rabbit. Am J Physiol Heart Circ Physiol 285:H173–H182

Takeshita S, Zheng LP, Brogi E, Kearney M, Pu LQ, Bunting S, Ferrara N, Symes JF, Isner JM (1994) Therapeutic angiogenesis: a single intraarterial bolus of vascular endothelial growth factor augments revascularization in a rabbit ischemic hind limb model. J Clin Invest 93:662–670

Tsutsumi N, Yonemitsu Y, Shikada Y, Onimaru M, Tanii M, Okano S, Kaneko K, Hasegawa M, Hashizume M, Maehara Y, Sueishi K (2004) Essential role of PDGFRα-p70S6K signaling in mesenchymal cells during therapeutic and tumor angiogenesis in vivo: role of PDGFRα during angiogenesis. Circ Res 94:1186–1194

Weitz JI, Byrne J, Clagett GP, Farkouh ME, Porter JM, Sackett DL, Strandness DE Jr, Taylor LM (1996) Diagnosis and treatment of chronic arterial insufficiency of the lower extremities: a critical review. Circulation 94:3026–3049

Yonemitsu Y, Kitson C, Ferrari S, Farley R, Griesenbach U, Judd D, Steel R, Scheid P, Zhu J, Jeffery PK, Kato A, Hasan MK, Nagai Y, Masaki I, Fukumura M, Hasegawa M, Geddes DM, Alton EW (2000) Efficient gene transfer to airway epithelium using recombinant Sendai virus. Nat Biotechnol 18:970–973

Yonemitsu Y, Matsumoto T, Itoh H, Okazaki J, Uchiyama M, Yoshida K, Onimaru M, Onohara T, Inoguchi H, Kyuragi R, Shimokawa M, Ban H, Tanaka M, Inoue M, Shu T, Hasegawa M, Nakanishi Y, Maehara Y (2013) DVC1-0101 to treat peripheral arterial disease: a phase I/IIa open-label dose-escalation clinical trial. Mol Ther 21:707–714

Index

Y. Nagai (ed.), *Sendai Virus Vector: Advantages and Applications*,
DOI 10.1007/978-4-431-54556-9, © Springer Japan 2013

R

MIX
Papier aus verantwortungsvollen Quellen
Paper from responsible sources
FSC® C105338

If you have any concerns about our products,
you can contact us on
ProductSafety@springernature.com

In case Publisher is established outside the EU,
the EU authorized representative is:
Springer Nature Customer Service Center GmbH
Europaplatz 3, 69115 Heidelberg, Germany

Printed by Libri Plureos GmbH
in Hamburg, Germany